INTRODUCTION TO
CONTEXTUAL
PROCESSING
THEORY AND APPLICATIONS

INTRODUCTION TO CONTEXTUAL PROCESSING

THEORY AND APPLICATIONS

Gregory L. Vert, S. Sitharama Iyengar, and Vir V. Phoha

CRC Press
Taylor & Francis Group
Boca Raton London New York

CRC Press is an imprint of the
Taylor & Francis Group an **informa** business

A CHAPMAN & HALL BOOK

Chapman & Hall/CRC
Taylor & Francis Group
6000 Broken Sound Parkway NW, Suite 300
Boca Raton, FL 33487-2742

Printed in the United States of America on acid-free paper
10 9 8 7 6 5 4 3 2 1

International Standard Book Number: 978-1-4398-3468-8 (Hardback)

Library of Congress Cataloging-in-Publication Data

Vert, Gregory.
    Introduction to contextual processing : theory and applications / Gregory Vert, S. Sitharama Iyengar, Vir V. Phoha.
        p. cm.
    Includes bibliographical references and index.
    ISBN 978-1-4398-3468-8 (hardback)
    1. Semantic computing. 2. Context-aware computing. I. Iyengar, S. S. (Sundararaja S.) II. Phoha, Vir V. III. Title.

QA76.5913.V47 2010
004.6'5--dc22
                                                        2010042054

Visit the Taylor & Francis Web site at
http://www.taylorandfrancis.com

and the CRC Press Web site at
http://www.crcpress.com

*To my colleagues for their support in the creation of
this book. Additionally, my deep gratitude to Dr. Petry,
Dr. Morris, Dr. Serroa, my family, and my advisors
Dr. Stock and Dr. Frincke at the University of Idaho.*

---

*Scholars are a product of their communities,
associations, and peers. We are changed forever by
our fleeting associations with great minds and take
a little bit of everyone with us when we leave.*

Gregory L. Vert

*I want to dedicate this book to Professor Thomas
Kailath (Stanford University) and Sarah Kailath for their
wonderful contributions to science and technology, and
their kindness to all of us during the last fifty years.*

S. S. Iyengar

*I dedicate this book to Li, Shiela,
Rekha, Krishan, and Vivek.*

Vir V. Phoha

Table of Contents

Preface

Contextual computing has been around for several years with a variety of names such as *pervasive computation* and *omnipresent computing*. Recently there has been a drive toward making personal digital assistants (PDAs) more aware of their environment. For example, a cell phone may sense that it is in a conference room and reason that it should divert calls to voice mail. In conjunction with this trend, there has also been an equally significant trend toward peer-to-peer (P2P) distribution of information, in which an unprecedented number of people can have access to information.

While considering these trends, the concept of context-sensitive computation developed. The initial idea was that information would drive the type of processing that was done on it instead of the traditional model of systems and software being static in the way they process information. The idea for this book was born from merging this concept with those of pervasive computation, peer-based hyper–data distribution, and global access to information. Thus the idea of contextual computation was born.

To the best of our knowledge this idea is seminal, borrowing from a number of areas of computer science including sensors, information systems, logic, security, graphics, networks, and mathematics. This book presents a comprehensive model for how a contextually based processing system might be constructed. It discusses the components of such a system, the interactions of the components, key mathematical foundations behind the model, and brand-new concepts that are necessary to make such a system operate.

Because the idea of contextually driven processing is comprehensive, it is a very large idea. Anyone from a developer to a company to researchers and government entities may want to borrow parts of the model or design entire systems based on the concepts developed in this book.

Additionally, the model presented is hyperscalable. Originally it was envisioned that the model would be utilized to manage manmade and natural disasters. Chapter 1 sets the stage for this discussion. However, upon further refinement of the concept, it was realized that the model being developed was a *global model* for the sharing of information among governments and worldwide entities. Looking at the opposite end of the spectrum, the model and its principles could be scaled down to implementation in small groups of people, within corporations, at the state level for emergency response, and at the national level. The details of implementation will vary based on the scale but should have all of the principles and components of the model presented in this book.

As a brief overview of the chapters, Chapter 1 starts with the original vision and motivation for the context-based processing model. It was conceived initially to be utilized for disaster management. This chapter starts laying the groundwork for defining the key *dimensions* of a model for contextual processing.

Chapter 2 is a key chapter. It starts with an examination of what data are and all the ways that data can be characterized and described. It is important to understand data intimately, because in subsequent chapters a context model is developed that can manage *any* type of data ever created or that will be created in the future. With an understanding of data, the definition and data structure for a context are developed. The concept of multiple contexts about the same event is presented, and the issues of conflicting information, missing information, or unknown information in multiple event-related contexts are discussed. The argument for aggregation of contextual information is made as a way to build more perfect knowledge from imperfect data collection. In order to aggregate, one must be able to reason about similarity among context data vectors with disparate data types. Some standard methods for determination of similarity in contexts are reviewed and discussed as criteria for determination of similarity and aggregation into a supercontext. This chapter then continues by developing a semantic model for contexts that is the basis for their processing. A grammar using the semantic model is then developed that can produce language-driven, context-specific processing actions. The model allows the implementation by a given entity (government) to map system-specific processing actions onto the language and thus has an open architecture. Toward the end of this chapter, the paradigm of a *global context* is examined and the argument that a comprehensive model cannot exist for political and legal reasons is presented. This leads to the notion that a global *core contextual model* could be developed with entity-specific extensions. Finally, this chapter concludes with a discussion about the quality metrics one might want to develop to describe the quality of a context's data. This is a key notion because the quality of contextual information is related to the confidence of the information and thus its application, processing, and transformation into knowledge.

Chapter 3, with a contribution from M. D. Karim, continues onward with reasoning about similarity among contexts and becomes a more formal introduction to rigorous mathematical methods that can be applied to contexts. It builds on the similarity analysis methods that are introduced in Chapter 2.

With the notion of the globality of contextual information construction comes the need for fusion of information. Chapter 4, containing contributions from N. P. Kavya and N. C. Sathyanaray, starts to consider in detail how contextual information may be fused. The architectural model draws the parallel that a context can be derived from multiple sources and that this model has similarities to that of how a sensor network collects and fuses information. This chapter examines basic sensor data fusion theory and then applies it to the contextual fusion of information.

With contextual modeling defined and the ability to aggregate by similarity and logic, the contextual information must be disseminated to a world of consumers who may be unaware that the information exists. Chapter 5, with a contribution from Jean Gourd, discusses how consumers of contextual information may be discovered by association with primary known consumers. It then examines similarity in consumption needs via the use of similarity ontologies. Finally, mechanisms for routing contextual information to consumers are examined against the current capabilities of the Internet. The streaming, high-bandwidth requirements of contextual data processing are discussed.

Chapter 6 examines the need for a new type of model for a data repository to manage contextual data. Due to the complexities of data types and selection ambiguity problems for the retrieval of contextual information, and their rules and contextual security values, a new type of paradigm is developed that can manage contexts and the dimensions of contexts. This model has the capability to manage any type of contextual data existing or created in the future by making the argument that less granular methods of management are actually more powerful than current approaches.

The final chapter on security of contextual data, Chapter 7, examines the needs and requirements of contextual security in a global environment.

This chapter looks at traditional methods and models of security which are hierarchical in nature. This chapter develops and proposes a new, open-architected model for doing security where the security flows with a context and is contextually related. This model mathematically relates the dimensions of contextual information and makes a comment about the level of security that might be required. This security model can be called *relational thematic pretty good security* (RTPGS). It provides a relativistic-based method for comparing the levels of security on a supercontext, and it leaves open the application of existing techniques. The model introduces the idea of a *brane* in the determination of security level. The brane mathematical surface has the capability of integrating the dimensions of time, space, impact, and similarity into a model for global security on a context.

Each chapter starts with a section called "Theme." These sections should be read first because they will describe ideas that may be unfamiliar but necessary to understand the topic of the chapter.

This book is meant to propose and describe a new viewpoint on global information systems operation. As such, it is conceptual in nature. A myriad of research questions could be investigated by almost all disciplines of computer science on the topic of contextual processing. Successful research in the area could provide the bedrock of a valuable paradigm shift in the next generation of IT systems design and processing of knowledge.

Gregory Vert

Center for Secure Cyberspace, Louisiana State University
Texas A&M University, Central Texas

About the Authors

Gregory Vert, CISSP, received his undergraduate degree with a specialization in geographic information systems (GIS) from the University of Washington, a master's degree from Seattle Pacific University in information systems management, and a PhD in computer science from the University of Idaho. While at the University of Idaho, he developed dissertation work in visualization methods for state changes in systems under attack and in advanced database models and methods.

Dr. Vert's research and publication record are diverse and extensive. He investigates and publishes about four different major research tracks. The first track centers on the development of advanced methods for intrusion detection and autonomous system response. This track includes visualization of state changes in a system as it is attacked. His second research area is in the development of advanced data management models that can store any type of information that currently exists or will be created in the future. This work centers on the application of set modeling. He also works in biometrics and bioinformatics. His most recent research effort was the development and integration of the new field of contextual processing. This work seeks to integrate current models of the semantic web and cloud computing with new original models into a new powerful paradigm for advanced information management. Such a model could become the basis for a future part of the Internet and advanced designs for information processing

Dr. Vert has worked in industry for 14 years for companies that include IBM, American Express, and Boeing. While at American Express he co-designed a portion of their worldwide database system. He has taught computer science at the University of Idaho, Portland State University, and Texas A&M. He has graduated 18 Masters students, mentored 1 PhD student, and published more than 45 conferences and journal papers.

Dr. S. S. Iyengar is the chairman and Roy Paul Daniels Chaired Professor of Computer Science at Louisiana State University, Baton Rouge, as well as the chaired professor at various institutions around the world. His publications include 6 textbooks, 5 edited books, and more than 380 research papers. His research interests include high-performance algorithms, data structures, sensor fusion, data mining, and intelligent systems. He is a world-class expert in computational aspects of software engineering, algorithms, and other aspects of computer science.

He is a Fellow of IEEE, ACM, AAAS, and SDPS. He is a recipient of IEEE awards, best paper awards, the Distinguished Alumnus award of the Indian Institute of Science, Bangalore, and others. He has served as the editor of several IEEE journals and is the founding editor-in-chief of the *International Journal of Distributed Sensor Networks*.

His research has been funded by the National Science Foundation (NSF), Defense Advanced Research Projects Agency (DARPA), Multi-University Research Initiative (MURI Program), Office of Naval Research (ONR), Department of Energy/Oak Ridge National Laboratory (DOE/ORNL), Naval Research Laboratory (NRL), National Aeronautics and Space Administration (NASA), U.S. Army Research Office (URO), and various state agencies and companies. He has served on U. S. National Science Foundation and National Institute of Health panels to review proposals in various aspects of computational science and has been involved as an external evaluator (ABET-accreditation) for several computer science and engineering departments.

Dr. Iyengar had 39 doctoral students under his supervision and the legacy of these students can be seen in prestigious laboratories (JPL, Oak Ridge

National Lab, Los Alamos National Lab, Naval Research Lab) and universities around the world. He has been the program chairman of various international conferences.

Dr. Vir V. Phoha is a professor of computer science in the College of Engineering and Science at Louisiana Tech University. Professor Phoha received an MS and a PhD in Computer Science from Texas Tech University. He holds the W. W. Chew Endowed Professorship at Louisiana Tech and is director of the Center for Secure Cyberspace. He has won various distinctions including Elected Fellow (2010), Society of Design and Process Sciences (SDPS); ACM Distinguished Scientist, 2008; IEEE Region 5 Outstanding Engineering Educator Runner-Up award, 2008; Research commemoration awards at Louisiana Tech University (2002, 2006, 2007, 2008); Outstanding Research Faculty and Faculty Circle of Excellence award at Northeastern State University, Oklahoma; and, as a student, was awarded the President's Gold medal for academic distinction.

Professor Phoha's most recent book is *Sensor Network Programming* (Wiley 2010). He has received funding from the National Science Foundation, Office of Naval Research, Army Research Office, Air Force Office of Scientific Research, Air Force Research Laboratory, and Louisiana Board of Regents, among others. He is a PI currently of active grants of more than $9.5 million and a Co-PI of grants of more than $6 million.

He has done fundamental and applied work in anomaly detection in network systems, in particular in the detection of rare events. He has 8 patent applications and many reports of inventions. He is author of more than 90 publications and author/editor of 5 books: *Fundamentals of Sensor Network Programming: Applications and Technology*, Wiley (2010); *Introduction to Contextual Processing: Theory and Applications*, CRC Press (2010); *Quantitative Measure for Discrete Event Supervisory Control*, Springer (2005); *Internet Security Dictionary*, Springer-Verlag (2002); and *Foundations of Wavelet Networks and Applications*, CRC Press/Chapman Hall (2002).

Contributors

Dr. Jean Gourd
Department of Computer Science
Louisiana Tech University
Ruston, Louisiana

Dr. N. P. Kavya Naveen
Department of Computer Science
    and Engineering
RNS Institute of Technology
Bangalore, India

Dr. Md. E. Karim
Center for Secure Cyberspace
Louisiana Tech University
Ruston, Louisiana

Dr. M. V. Sathyanarayana
Malnad College of Engineering
Hassan, India

CHAPTER 1

The Case for Contextually Driven Computation

THEME

As the second millennium has been dawning, there has been a remarkable shift in the computing paradigm away from the concepts of hardware processing data in a structured monotonic fashion. This evolution has become increasingly spurned on by some of the most spectacular natural and manmade disasters our civilization has ever seen. Among such disasters are September 11, tsunamis in Asia, volcanic eruptions, and catastrophic nuclear accidents. In all of these events, there has been a structure of information distribution, usage, and processing that has not kept up with the needs for information content and a context to determine how information will be processed. Examples of this are information not being shared among countries and entities, uncorrelated inferences of meaning, and criticalities of information processing in a fashion that truly does not serve various perspectives' needs. If these information resources could somehow be integrated more completely and perhaps in an automated fashion based on some paradigm or model, they might be much more effectively utilized. Such a paradigm might consider the environment that information is collected in as a means to determine the type of processing it receives. This type of concept might be referred to as *contextual processing* (CP).

Contextually driven processing would be a new paradigm driven by the environment and semantics of meaning of an event and information about the event found in the environment. This type of processing would require a context which might contain metadata about the event's data. If the event was a natural or manmade disaster, such metadata will most likely have a spatial and temporal component. This type of data can be very complicated to manage and rich in information content, and thus a good subject for inclusion in a contextual-processing model. We will explore this notion in more detail in subsequent chapters. Key to the concept of contextual processing is that the environmental, spatial, and temporal information surrounding the event should dictate how information is stored, processed, and disseminated.

When considering the development of a contextual-processing model, which may operate at a global scale potentially, one has to examine the notion of security for such a model. Complicating this need has been the increasingly hostile threat environment of the Internet. If the way information is processed is going to be changed, then new models of how it will be secured will also need to be developed. Traditional methods for mitigating threat have been well studied. Increasingly, the term describing all aspects of information integrity has been referred to as *information assurance* (IA). IA is a much broader definition than the traditional definition of what composes computer security. IA has now come to represent the comprehensive approach to information integrity within an entity such as a division of the government or a corporation. IA considers and addresses the policy, legal, human factor, hardware, software, and investigational forensics of information handling. However, IA is typically applied to a centralized domain-specific entity where information is the sole property of its owning organization. Ownership leads to the traditional view that information is secondary to how it is processed and how it is secured. This perception and understanding also create islands of security centered on organization in the scheme of the World Wide Web. With global distribution and contextual processing, IA in contextual-driven processing becomes a complicated question. In order to have a contextually based processing system, one must define IA not based on the simplicity of having a domain of control but upon larger issues. The question becomes in actuality how to have security and integrity in information flow where no one really has central authority to implement it. In simpler terms, in an environment of hyperdistributed information sharing, context-specific processing, political boundaries, and conflicting legal systems which drive regional security policy, how does one implement peer-based IA mechanisms, implementations, and policies that effectively combat the threat environment of the Internet?

The need for, and issues of, contextual-based processing can be illustrated by considering two recent disasters: Three Mile Island and the Indian Ocean tsunami. These examples can demonstrate the need to interrelate information scattered across multiple entities and to process it based on the environmental context of the information. This is where the concept of information being the primary player leads to the concept of information flow to the appropriate resources, where it should dictate a context for processing its analysis.

In order to illustrate the above points, it is useful to examine two different natural and manmade disasters most people are familiar with, in which a concept shift on how information is processed could have remediated the situation. In order to contrast the events and the failures of traditional information-processing approaches, there is a need to develop a qualitative scheme for comparing and contrasting the events. This scheme begins to hint at some of the classes of information that should be part of a global contextual-based processing model. For contrast and comparison purposes, we will define the following attributes that can be utilized to characterize an event:

- *Temporality*: The time period in which an event unfolded from initiation to conclusion (medium, short, long, or null)

- *Damage*: The relative damage caused by the event in terms of both casualties and monetary loss (mild, none, moderate, severe, or catastrophic)

- *Spatial impact*: The spatial extent, regionally, in which the event occurs (point, none, small, medium, or extensive)

- *Policy impact*: Directly driving the development of IA (security) policy both within a country and among countries (a statement of how policy was affected as a result of the event)

In the following events, the result of information being processed in the traditional methods will be examined, aspects that support the argument for context-based information processing will be presented, and finally some of the considerations of IA on context-sensitive processed information will be discussed.

1.1 THE THREE MILE ISLAND NUCLEAR DISASTER

The Three Mile Island (TMI) nuclear reactor is located in Dauphin County, Pennsylvania, near the town of Harrisburg. On March 28, 1979, TMI suffered a partial core meltdown in Unit 2, a pressurized water reactor. The TMI meltdown was significant because people were concerned about nuclear power and its safety due to a movie shown a few weeks before, *The China Syndrome*, which dramatized a meltdown. What was also significant was the lack of understanding about what was actually occurring in the reactor vessel because there was very little sensing equipment to measure water levels, external radiation, and generalized operational information, for example, no cameras monitoring the core.

There were about twenty-five thousand people living in the immediate area of the site at the time of the incident, which started just after 4:00 p.m. [1]. Metropolitan Edison, the company that operated the reactor, had Vice President for Power Generation Jack Herbein erroneously minimize the event by describing it as a "normal aberration." In reality, the operators of TMI had little understanding of what was actually occurring in the reactor as the cooling capacity in the core region of the reactor started to fail. This led to an increasing buildup of heat that could in theory melt through the reactor vessel. Operators did not comprehend the scope of the threat due to the poor information then coming from inadequate sensors and monitoring. At around 7:00 a.m. the next day, a Site Area Emergency was declared, and at 7:24 a.m. this warning level was upgraded to a General Emergency, the highest U.S. Nuclear Regulatory Commission (NRC) accident level. It was not until thirty-five minutes later that the local radio station announced a problem at TMI and the public first became aware that something was very wrong.

Even five days later, the scope of the problem was still not clear and federal and local authorities were unable to decide whether to evacuate the general surrounding areas. What was even more concerning was the way in which the plant was brought under control by venting radioactive steam (to relieve pressure in the reactor) straight into the atmosphere without notifying the public. It was estimated that around 13 million curies of radiation were released over the period of a week. At the end of the event, the reactor was brought under control, but people had been exposed to radiation, the true impact of which is still not known; and the full details of what had actually happened, the melting of the core, were not discovered until much later. It could have been much worse.

A contextual-processing paradigm might have allowed for the *integration and processing* of information about this event in a faster and more knowledgeable fashion. From the perspective of traditional information-processing methods, this disaster has several striking and interesting characteristics. First, information was largely centered in the control room at TMI; this information was not readily and fully available to decision makers at the NRC and the public. Obviously, one of the concerns about dissemination of information to the public was panic. The other problem was that the information in a number of cases was actually not being sensed or was being sensed erroneously. For instance, cameras in the reactor vessel and better instrumentation of the vessel could have provided richer information content and led to faster and better decision making. In the context-sensitive paradigm, data might have been aggregated to mitigate sensor ambiguity. Contextual consideration of the spatial and temporal evaluation of the existing data might have led to different analysis and processing. As an example, control instrumentation might have been evaluated with construction data and analyzed in real-time in conjunction with NRC decision-making processes.

A relative *characterization* of the data on events at TMI, considering their context, might be as follows:

Temporality: Medium time frame

Damage: Mild, evacuations, and no loss of life

Spatial impact: Moderate, 3–4 miles

Policy impact: Policy development, antinuclear sentiment, and energy dependence on oil

1.2 INDIAN OCEAN TSUNAMI DISASTER

On December 26, 2004, an earthquake with an estimated magnitude of between 9.1 and 9.3 occurred in the Indian Ocean. It was the second largest earthquake ever recorded. The approximate time duration was between eight and ten minutes; this is unusually long for an earthquake, most of which last less than a minute or two. The quake caused the earth to shake an estimated 0.5 inches off its axis. The earthquake was so powerful that it caused other earthquakes to occur as far away as Alaska. This quake, referred to as the Sumatra earthquake, caused countless damage and loss of life.

The quake originated off the west coast of Sumatra in Indonesia and triggered a number of very large and dangerous tsunamis. These tsunamis hit most of the shorelines around the Indian Ocean without warning of any kind. The result was 225,000 people dead in multiple countries, countless injuries, and coastal flooding with tsunami waves reaching close to a hundred feet. The hardest-hit countries were Indonesia and Thailand, with lesser damage in India and other countries. The devastation prompted a humanitarian effort from the worldwide community. In total, $7 billion was spent on recovery and mitigation of the disaster.

This disaster highlighted the importance of having the *right information at the right moment* with the correct granularity to make accurate decisions. This is something that contextual processing might address by the integration of information and its correct dissemination and analysis. In the tsunami event, information was disjointed, disparate, and not disseminated such that more effective decision making could occur. With adequate sensors, satellite imagery, observational information, and other data collection sources, contextual data could be integrated and disseminated, providing better warning to governments and cities surrounding the Indian Ocean which might have saved countless lives. However, such informational knowledge was not available because there was not a context model to drive processing and dissemination of knowledge. In such a model, there is a need for accuracy, analysis, and prediction of the severity of a projected tsunami so as not to create undue panic in the general population. Context-based dissemination of such information also comes with the requirements of security measures to protect and authenticate such information to guarantee its integrity, fidelity, and accuracy. None of this capability was in place on the day of the tsunami because only limited, traditionally based processing functionality was being utilized.

A relative contextual characterization of these events based on the previously presented categories might be as follows:

Temporality: Medium time frame

Damage: Huge

Spatial impact: Large

Policy impact: Low, cooperation in the creation of a warning system

1.3 CONTEXTUAL INFORMATION PROCESSING (CIP) OF DISASTER DATA

The previously discussed events may be characterized as disparate natural (tsunami) and manmade (TMI) events that a different type of information-processing model could have helped in analyzing data, predicting outcomes, disseminating information, and thus mitigating the effects of the event. The paradigm describing this can be referred to as the *context-based processing model*. This model implies that information should flow, or disseminate, to where it is needed and that the semantics of an event, collected and integrated from many sources, can determine the type of processing, analysis, and correlation methods that the information receives and thus the knowledge derived. Context-based processing of events may be applied to the spectrum of processing activities as raw data are turned into knowledge and potentially actions. Some of these activities include the selection of computational resources to process an event's data, the analysis method applied, the correlated knowledge discovery, the means and techniques of dissemination, visualization, and the derivation of implications and perspective of meaning for a given event. This type of processing model can be very heterogeneous in the type and format of data being transformed. The disparate data can range from binary data to images and tabular, audio, visual, and textual written descriptions. The range of data is due to the fact that the context model seeks to integrate data and derive meaning from a wide variety of sources, including sensors, humans, and knowledge discovery methods.

As alluded to earlier when characterizing events, a context-driven processing model will have the complicating environmental factors of geospatial and temporal elements. The geospatial domain means that information is collected and stored at a distance from where it may be processed and used for making a decision. This means that CIP processing must have a comprehensive model to route information based on semantic content to the appropriate processing location and dissemination channels. CIP processing also has a temporal component. It can be collected over periods at regular or irregular intervals, and the time when the information is collected also may determine where the information is sent and the context of how the information is processed. For instance, information that is collected for simple monitoring may, in the case of the tsunami, flow to research institutions around the world for storage and analysis at some point in the future. In contrast, noticing earthquakes on the ocean floor may route collected information to countries surrounding an ocean for immediate high-speed analysis, critical real-time decision making, and rapid dissemination. Processing, in this case based on the *immediacy* of threat, may contextually shift computational resources away from routine tasks in order to handle high-priority emergency processing based on the context model. Some factors that can be described in CIP processing can be referred to as *information criticality factors* (ICF), where the concept of *criticality* determines the need for a context shift in processing by computational resources. These factors become key to thinking about how a context-based processing model might operate. They may include but are not limited to the following:

- Time period of information collection

- Criticality of importance

- Impact, for example, on financial data and cost to humans

- Ancillary damage

- Spatial extent

- Spatial proximity to population centers

The ICF and potentially others could be used to evaluate threat, damage, and criticality of operational analysis. Other factors affecting CBI processing might be based on the *quality* of the data by characteristics such as the following:

- *Currency*: How recently were the data collected, are the data stale, and do they smell bad?

- *Ambiguity*: When things are not clear-cut—for example, does a 1°C rise in water temperature really mean global warming?

- *Contradiction*: What does it really mean when conflicting information comes in different sources?

- *Truth*: How do we know this is really the truth and not an aberration?

- *Confidence*: Does the information reflect the truth about an event?

In order to develop a global, contextually based processing model, two key concepts need to be architected into the model. The first is the information infrastructure, and the second is the IA model for contextually based information flowing around the Internet in a relatively flat model to different cultures, legal frameworks, and security systems.

An information infrastructure, like the lanes of a freeway, is utilized in determination of where data flow and becomes an information model about the informational relationships among producers of contextual information and consumers of that information. Questions of trust, strong connections, and weak connections need to be considered in the determination and definitions of infrastructure and can become the basis for derivation of an IA model.

1.4 CONTEXTUAL INFORMATION PROCESSING AND INFORMATION ASSURANCE (CIPIA) OF DISASTER DATA

IA and security in a CIP architectural model will have to be very different from current approaches and should be context sensitive. This is because such a system would operate at a global scale spanning different organizations, entities, and countries with widely different policies on use, privacy, and legal systems In context-based processing, IA would be a highly distributed, granular, open-architected, flat model. This is a different type of approach than the traditional, hierarchical, centralized approach to current IA.

In the current environment, security and integrity functions are centralized within organizations and entities by a controlling authority. This traditional model makes development of IA policy and its implementation relatively simple because a small group of people administer and operate a system. However, due to the global geospatial- and temporal-based aspects of CIP processing, CIPIA would have to operate as a function of context. In short, a global contextual-based system would be an information system model for the world; therefore, its IA measures would need to be globally defined, open architected, and not based upon geopolitical entities. Some goals of the CIPIA model might be to guarantee the following:

- It provides a relativistic indicator of which information is more important to secure.

- It supports open architectures where the application of existing methods can be integrated.

- Comparison of security requirements is standardized in a fashion such that limited computational resources can be applied to the most pressing IA requirements.

- Information arrives at its destination and has integrity.

- Access to the information can be controlled.

In order to implement CIPIA, standard approaches to IA could be applied (e.g., encryption and authentication) but in a new fashion. Such a model might seek to define or develop the following:

- New rapid methods for authentication of geospatial and temporal information

- Dissemination models to people with a need to know, perhaps through the modeling of social networks

- Standardized models and methods for evaluation of the security level of contextual data in a way that they can be compared and quantified

- Access control such that worldwide users and organizations can grant access in a standardized fashion

- New media-neutral, robust integrity measures for verifying that contextual data are genuine, similar to the concept of watermarking images, audio information, and textual information

- Development of a security model that can drive interorganizational and international security policy development in a standardized fashion for a multinational setting considering disparate legal and privacy requirements

- Political- and legal-sharing understandings modeled on standardized levels of authorization for access to information

- Insider threat controls and checks such that a single person cannot trigger panic through the dissemination of erroneous information

The above ideally needs to be embedded with contextual information as it flows across the Internet, and needs to be enforced in the context of the *type of the information* when it is processed.

From the above discussion, the first key component of the notion of contextual processing can be defined: characterizing the dimensions. A *contextual dimension* in this case is a key attribute of contextual processing that drives the definition of its architecture and IA model. Key dimensions are thought to be the core minimum that is required to define an event in the contextual model. The key dimensions derived from the above discussion are as follows:

- Temporality
- Spatial extent
- Impact, which drives policy and IA development and security levels
- Similarity, which can allow reduction in computational overhead

The above are the initial tenets that any model for contextual processing should and must consider in order to effectively deal with natural or man-made disasters.

1.5 COMPONENTS OF TRADITIONAL INFORMATION TECHNOLOGY (IT) ARCHITECTURES

Having examined the need for contextual processing and some of the issues, we next turn to how it might be utilized in a global information system. Almost any large IT system has the following architectural elements which comprise the basis for the selection of subsequent chapter topics in this book. The components are as follows:

- A repository for information
- The capability to transmit data with security and integrity
- An ability to derive knowledge from the process of fusion of data
- The ability to reason about information
- A capability to disseminate information broadly

When one considers IT systems architecture, all or most of the above elements are included in some shape or form. However, current architectural components of an IT system are not as powerful as they could be when contextual processing concepts are applied to their design and operation.

1.6 EXAMPLE OF TRADITIONAL IT ARCHITECTURES AND THEIR LIMITATIONS

To illustrate the operation of current architectural designs for IT, we present and discuss the Indian Ocean tsunami case example and how the components of IT systems mentioned previously may typically operate.

The first assumption that needs to be made in this example is that data collection capabilities exist that can collect information about changes in the ocean around the origins and peripheries of the tsunami. Such capabilities may include buoys, ocean sensors, satellite imagery, ships at sea, and human-based observation. We shall refer to all of these entities as *event objects* and denote them as eo_i, where i is merely a unique identifier of the specific event object data collection point. The second assumption to be made is that most of these event objects, if not all of them, are unaware of each other and most probably not interoperable. At the moment the tsunami occurs, there is plenty of information that could be collected to clearly identify the event. However, pathways for massive distribution and dissemination of this information may not and probably do not exist. This is because a buoy sensor system may be operated by a government agency, whereas the satellite imagery and remote-sensing data from space may be heading to its owning agency, perhaps NASA on the other side of the world. Because of the fragmentation of information found in this scenario, there can be limited dissemination of information about the event which might be of interest to almost any person or entity along the shores of the Indian Ocean; This could amount to literally millions of people. The problem of distribution of information to entities that might be interested in such information is a very real problem in *current* IT systems designs.

Given that hyperdistribution of information collected from event objects could exist as a capability, there is a need for management of the information so that it can be analyzed efficiently. Currently, this is largely done with monolithic relational databases. Relational technology was invented in the 1970s and accordingly has limitations inherent in its dated theoretical basis. In particular, data must be organized into relations, which is easy to do for data that are clearly understood and known about during the design process. The problem of a priori knowledge limits the application of relational technology to contextual processing because new types and structures of data are being created constantly. Additionally, relational technology is not particularly well suited to the storage of spatial and temporal data due to the complex types of relationships found in such data. Typically, contextual data will have the same types of properties and relationships as spatial and temporal data, and thus need a new type of data management model that is not rigid in its approach and can maintain associations among disparate, extremely diverse, and unlimited types of data formats. Finally, limitations on retrieval methods, especially *Structured Query Language* (SQL) capabilities, may make it difficult or even nearly impossible to locate all the data required for contextual and semantic interpretation about an event. Also, the perpetual problem of not knowing what data you want to retrieve for analysis could be critical to contextual and semantic knowledge derivation processes.

If one assumes that massive hyperdistribution of information is a capability in a current system and advanced data management models are employed, this leads to the next architectural limitation of current systems, *the ability to reason about information* and particularly its similarity and contextual-based semantic meaning.

This problem exists because the data that an event object may be collecting can have any one of the following classes of problems:

- Conflicting data

- No data

- Limited data

- Ambiguous meaning in data collected

Since raw data in conjunction with contextual environmental data about an event object are the key components in the derivation of contextual knowledge and thus processing and response, the above creates problems. In contextual processing and especially reasoning, some of the data or information may indicate by its semantic interpretation (context) that a tsunami is occurring, and some may not seem to have any particular semantic meaning. Current approaches to dealing with ambiguity in information are fairly limited when you consider that most do not attempt to reason about the context of the information as a means to determine that all the information collected has the same semantic meaning. For instance, a wave in the ocean may mean *tsunami* or *not a tsunami*. A wave in the ocean in conjunction with an earthquake and circular expanding temporary increases in water pressure at certain sensors probably does mean *tsunami* and can be inferred using contextual information about the event. Logics that attempt to process within the framework of context typically do not exist in current IT systems but could be very relevant in contextually based IT designs.

If one assumes that current IT has hyperdistribution models and capabilities, reasoning about similarity of contexts for semantic interpretation and a repository that is powerful enough to support the advanced information management requirements of these systems, there is still a need to derive *knowledge based on context* from all the information. This again is not typically a capability of current IT systems. In this case, the contextual model could be thought of as a black box where data and contextual information enter, and fusion works in conjunction with context to produce actionable knowledge (K). K can then be an input to another black box with further contextual information producing as output contextually based processing rules (R). Currently data fusion is done post facto of data collection, and it is done typically on isolated bodies of information. Because of the underlying mathematics and theory of data fusion and data mining, there is little if any consideration of context in the processing or interpretation of the result of such processes. There is currently no known paradigm in research or industry referred to as *contextual data fusion*; however, in a global contextually based processing system there is probably a need for exploration of current approaches and an extension of them to include contextual fusion and contextual data-mining methods. Contextual modeling may (1) determine how data are fused, (2) support the generation of better data sets for data mining, and (3) add semantic interpretation of results to produce a richer knowledge space and better response rules.

Finally, in the creation of a global contextually built IT system, there will be a need for data integrity, that is, a type of security. If we assume there exist perfect reasoning, storage, and dissemination mechanisms, there is a need for contextually based security mechanisms to assure the integrity of information and prevent maliciously induced mass hysteria. Current global IT systems are limited in their approaches and methods for securing data and knowledge. Typically, the large databases maintained about natural disasters will employ established and *dated* traditional security mechanisms, for example, logins and encryption. Spatial and temporal data tend to be voluminous and complex in structural semantics. On a global scale, large-scale transmission of voluminous spatial and temporal data is typically not done due to the intense computational and transmission overhead. If the information is sensitive and needs to be secured, the costs become even higher because the typical mechanisms for assuring integrity such as encryption can be extremely slow on large amounts of complicated data and therefore prohibitive. This is a limitation in current IT systems which also have hierarchically controlled security processes. In a global contextual system, data should be viewed as a flat model, almost like peer-to-peer (P2P) models where data flow to where they are needed, like amoebae. This vision requires advanced security models to assure the integrity of flowing information. Thus, there is a need to provide consumers of information with an indicator of the *level of security* that information might need that has meaning and relevance to the consumer for which they implement their interpretation of what that level implies a new concept that of *contextual security* based on relevance importance and relation to other data. With this notion comes the idea that *maybe not all data or data objects need to be secure*, but consumers do need to have reasonable assurances that the information they may take action on has integrity relevant to themselves. A mechanism not found in current IT, coined *pretty good security*, needs to exist in a contextually based global processing system. Such a model should include the principles and concepts that have been mentioned in this chapter.

Most current IT systems may have some limited degree or no degree of the following capabilities:

- A repository for information
- The capability to transmit data with security and integrity
- An ability to derive knowledge from the process of fusion of knowledge
- The ability to reason about information
- A capability to transmit and disseminate information

In order to develop thought and research on what contextually based global computation and processing may evolve into, the above components were deemed the initial key areas to define. The chapters included in this book were selected and organized to examine how the new model of contextual processing might be implemented and applied to the above architectural components. This work is not meant to be an exhaustive overview of every minutia in such systems; in fact, it is just the opposite. The ideas presented in this book will seek to present new paradigms, constraints, and capabilities that can integrate under a commonly defined model to produce the next generation of IT systems. The goal was not to dictate how such a system should be built but rather to encourage scholars and researchers to carve out their research niche and define global processing further according to their ideas. Graduate students are encouraged particularly to find ideas for thesis work, and to pursue and publish their ideas. It is hoped that this will set the stage for further thought on the subject, stimulate potential research, and paint a vision for global IT systems ten years into the future.

1.7 CONTEXTUAL PROCESSING AND THE SEMANTIC WEB

There is an important research effort going on involving development of the semantic web. The semantic web is somewhat similar to the paradigm of contextual processing. Web research is attempting, among other things, to model relationships among data and do this at a machine-readable level [2]. This involves the use of semantics and, to some extent, context. Contextual processing is somewhat similar in that it proposes to use context to do something else: control the processing of information. In this regard, contextual-processing research can benefit from the methods and techniques developed in semantic web research. Completion of the semantic web is still in the future, as research continues.

In contrast, contextual processing varies from the semantic web in several important and significant ways. First, it proposes the development of an entirely new infrastructure for IT. This infrastructure includes the key components of data management, new security models, knowledge grids, producer–consumer paradigms in the hyperdistribution of information, contextual logic, and contextual data mining. The models proposed are generally new and centered on the application of context. Additionally, in the contextual-processing model, information, its processing, and its derivation into knowledge would be controlled by semantics and contexts. Finally, the paradigm in the next chapter presents *classes of spatial and temporal data metaclassifiers* that can be utilized to drive the initially proposed semantic-processing grammar. It is possible that the development of such a system could be a good platform for the semantic web to run upon, and thus research should consider how these two concepts might integrate.

1.8 CONTEXTUAL PROCESSING AND CLOUD COMPUTING

Another area on which contextual-processing research might have impact is the notion of cloud computing. *Cloud computing* is defined to be a new supplement consumption and delivery model for IT [3]. It can be thought of as a paradigm where users consume services but don't have to actively manage the hardware that is providing the services. Thus it can be thought of as an abstraction in architectural design where infrastructure management is separate from application management.

There are several benefits to such an architectural model. The first of these is that smaller companies can compete on a larger scale but with less entry expense into a market, merely by purchasing or subscribing to necessary business services. Infrastructure utilization rates in a shared service system with many subscribers should in theory increase, providing improved efficiency of operation and delivery of advanced services. Such services can be characterized as providing broad access to the Internet and on-demand access that can be purchased on a subscribed basis.

While there may be a tendency to think of cloud computing as being represented by the Internet, it is actually more appropriate to perceive the Internet as being composed of N multiple existences of clouds tailored to the delivery of specialized services, created by specialized service providers and having loose affiliations and overlaps with other clouds in the universe of discourse.

Cloud computing is often confused with grid computing, utility computing, and autonomic computing, but it is really a distinct paradigm. *Grid computing* is a form of parallel processing among loosely affiliated systems that operate in unison to perform complicated computational operations. In a sense, this may be thought of as many smaller computers aggregating computational power to create a virtual supercomputer. In contrast, *utility computing* refers to the creation of metered or subscribed services and is similar to cloud computing. This type of operation can be demonstrated in online services that store user information. Finally, *autonomic computing* can be defined as computation that operates in a self-aware, self-organizing, and self-managing mode.

Cloud computing relies on services delivered through data centers built upon servers. This makes it architecture hierarchical in organization, at least at the levels where services are being generated and supplied. Most of these services are provided asynchronously and can be organized into a three-tier functionality. The lowest layer is referred to as the *infrastructure layer*, consisting of servers, networks, and storage media. The layer built on this is the *platform layer*, where services are generated. Finally, the *application layer* provides user and machine interfaces. The architecture is somewhat similar to the functionality found in the layered OSI model. Some goals of the cloud-computing layered model are that it should be self-monitoring, self-healing, and self-optimizing [3]. And key characteristics of such a system are described as its agility, reduced cost, hardware independence, resource-sharing capability, reliability, scalability, and security due to centralization of data.

Contextual processing has a role in cloud-computing research and some different focuses and goals. The first of these is the notion that contextual processing seeks to change the underlying systems' infrastructural designs and capabilities in order to create more powerful infrastructures. Thus, CP should be investigated for application to infrastructural-level cloud computing. This could lead to more powerful services because the infrastructural capabilities could become much more powerful as a result. Current approaches to cloud computing seek to organize existing technologies to provide more efficient utilization without addressing a fundamental *redesign* of current processes and hardware.

Contextual processing also varies from cloud computing in that it is designed to operate in a hyperdistributed environment where consumers of information are not aware of producers of services and useful information. In cloud computing, the notion of a cloud implies a flat management model of data and services when in reality the services are provided by a centralized source, thus following a traditional model for computation. The CP model seeks to extend research into a truly hyperdistributed peer-based global model for knowledge generation and consumption. In such a model, cloud computing should be investigated for how it would integrate into a flat management model.

Another difference between cloud computing and CP is that of security. Traditional thinking in security modeling sometimes reflects the thought that all data must be managed and thus secured by a centralized authority. However, from a logical standpoint it is impossible to manage *all* the data in the world at a single centralized location, as current attacks on the Pentagon and Internet demonstrate. This is a fundamental flaw in the way we currently think about security and also is computationally impossible. CP makes a different argument, and one that is much more pragmatic. Its argument is that what should be secured is based on the *relation* of that object and its data to other objects. While CP's initial approach to security presented in this book is based on the key defining dimensions of CP, it has been recently realized that the methods presented in Chapter 7 can be extended to manage the relationship of concepts and thought as a way to distinguish what is truly important to secure and what is not as important to secure.

Cloud computing does point out the problem of legal issues found with information sharing and data access, but does not really attempt to propose a new solution. This has prevented much more global implementation of the paradigm. CP proposes a possible solution: creating *core contexts* that are legal neutral and augmented with key extensions for communities of contextual processing. This is similar to a concept discussed in the next section about Universal Core (UCore). The key research area in core context is how to describe it and what it might be composed of. Perhaps core context, as discussed in the next chapters, would be a metaknowledge framework constructed to contain neutral information that could be shared globally or by communities of CP processors.

A similarity with cloud computing that CP strongly shares is the principle of asynchronous distribution and access to knowledge and contextual knowledge derivation. In the CP model, however, knowledge distribution would be envisioned to be at a much larger level than inside specialized computational clouds.

A goal of the contextual-processing model is to be a unifying and integrating framework for existing technologies and concepts, and to augment those technologies with the new approaches and concepts presented in this book. Cloud computing would be a technology that contextual-processing research should seek to integrate and understand its relationship with.

1.9 CONTEXTUAL PROCESSING AND UNIVERSAL CORE

Finally, another technology that is similar in some regard to contextual processing is referred to as *UCore*. This is a U.S. Government–sponsored project to facilitate sharing of data among government systems. UCore is based on extensible XML and seeks to define interfaces and interoperability between entities using government data. The need for UCore arises from the existence of disparate systems, processes, and infrastructure.

The UCore effort has interesting ideas that are somewhat similar to that of contextual processing. One of these is regarding the existence of knowledge domains and common vocabularies. UCore has developed specific extensions such as the C2 Command and Control extension and traces its roots to a U.S. Department of Defense (DoD) directive (8320.1) that proposed a very large data model for all DoD systems [4].

UCore can be implemented in two different ways: the first is full adoption of its concepts for information sharing, and the second is referred to as *vocabulary reuse*, which acknowledges that full implementation of

UCore may not always be possible. Some systems, particularly legacy or new systems, are going to be disposed to one or the other implementation method.

Finally, UCore does utilize a variety of standards to define itself. These include the Discovery Metadata Specification, which is used to locate resources of interest to UCore. UCore also utilizes an ontology-modeling language (OWL) and a markup language called Geographic Markup Language (GML).

Many elements of UCore should be investigated for extension and application into contextual processing. For instance, UCore only models one dimension of contextual data, that of spatial information. There are three other dimensions of contextual data (at present): those of similarity, impact, and temporality. Research could be done on GML to extend it into contextual-processing Temporal Markup Language (TML), contextual-processing Impact Modeling Language (IML), and contextual-processing Similarity Analysis Modeling (SAM).

Another key concept of UCore is that it seeks to increase interoperability and information exchange. The *set-based* model presented in Chapter 6 is a powerful new data repository and data-modeling paradigm that should be studied to see how integration of existing UCore data exchange interfaces could be applied. Set management methods could be integrated with UCore architectures to produce more power data management paradigms than currently exist.

Finally, UCore and contextual processing have a similar concept: a shared knowledge model with localized extensions of dialect. In CP, this is referred to as *core context* and *extensions*. UCore refers to this as knowledge domains and common vocabularies. UCore concepts are similar to CP but are a specialized subset of the CP vision and model because they focus on a specific application, DoD-specific data and information exchange. UCore does not focus on an open-architected global information exchange model such as contextual processing proposes. Additionally, UCore does not concern itself with hyperdistribution concepts or the *unknown consumer–unknown knowledge producer problem* addressed in Chapter 5. There is the potential for a considerable amount of research synergy that could be derived from the investigation of an integration of semantic web, UCore, and cloud-computing concepts into contextual processing, and the areas should not be seen as mutually exclusive.

1.10 THE CASE FOR CONTEXTUAL PROCESSING AND SUMMARY

In this chapter, an initial method of characterization of disasters was presented. These were utilized to review the TMI and Indian Ocean disasters and characterize them in a relative fashion. The weaknesses of current approaches to information processing were presented as a means to start building an understanding of how processing should change and lay the basis for the contextual-based processing model. After this analysis, key notions and concepts that a contextual-processing model might include were presented and discussed as a means to start thinking about how such a model would be different from current methods. Information assurance and security will be key to such a model if it is to function at a global scale or smaller scales. Issues key to IA were discussed and analyzed, and key components of IA in relation to the notion of context were suggested. A key notion of contextual processing, that of dimensionality being a central component to a model for context, was derived from an earlier discussion. The four central dimensions were identified that can uniquely define any contextual event and thus drive its processing, dissemination, and security. Finally, current IT systems architectures were examined and limitations were explained. These were contrasted with the operational vision of an advanced contextually based global-processing model.

The next chapter, Chapter 2, will build on the concepts and notions presented in this chapter that define a contextual-processing model. It will start by defining the development of contexts and how they might be aggregated by the dimension of *similarity*, then discuss processing semantics, contextual-processing grammars, and how processing may be layered onto a global model for context-driven event processing. Chapter 2 starts with a look at defining context, the idiosyncrasies, and classifications of data; semantic grammars can be developed that control processing without dictating how it is done and how similarities analysis might be conducted to mitigate missing or conflicting information defining a context. Chapter 3 addresses the IT architectural component, mentioned previously, of needing to reason about information and particularly the need to develop logics that can help determine if conflicting contextual information is really the same information. Chapter 4 introduces data fusion at a global scale and starts to set the stage for what will hopefully become an emerging research theme: data fusion based on context. Chapter 5 introduces thinking and framework about the hyperdistribution of information utilizing contextual ontologies. It presents the problem that consumers of information may not know that the information they want exists and that producers of such information may be unaware of the consumers that might want the information. Some thought and proposed methods are presented in this chapter for solving this problem. Chapter 6 presents a repository-neutral framework and model that addresses the advanced data management requirements of contextual data, its disparate nature, wildly varied data types, and the need to maintain relationships among large sets of data. Finally, Chapter 7 introduces some thinking about contextual security, which may emerge as a new area of security research. The notion that what needs to be secured may depend on the relationships a particular piece of information may have with another piece is presented. This concept cuts down on computational overhead for security on contextual data and has been given the name of *pretty good security.*

REFERENCES

1. Wikipedia. (N.d.). "Jack Herbein." http://en.wikipedia.org/wiki/Jack_Herbein
2. Wikipedia. (N.d.). "Semantic web." http://en.wikipedia.org/wiki/Semantic_Web
3. Wikipedia. (N.d.). "Cloud computing." http://en.wikipedia.org/wiki/Cloud_computing
4. Wikipedia. (N.d.). "UCore." http://en.wikipedia.org/wiki/UCore

CHAPTER 2

Defining the Transformation of Data to Contextual Knowledge

THEME

This chapter builds on the previous chapter and examines the nature of knowledge as it applies to natural disasters. The examination leads to the first key concepts in defining the new contextual-based processing model, that of dimension. The chapter continues with a detailed examination of the nature of data and information so that the more subtle nuances of the model may be understood in the similarity and analysis discussion. This is presented as a further defining basis for the context model. The initial model is then discussed and explained, and a potential semantic grammar is developed for controlling processing and dissemination of contextual information. This grammar is open architected in the types of specific application responses a user of the context model might apply to knowledge derived from contextual processing. The initial context model has some weakness because of how contextually based information might be collected. Therefore, the argument for aggregation of contextual data into supercontexts is examined. Aggregation is done via similarity analysis. Some potential methods are discussed for reasoning about similarity, and some ideas about how they could be applied to contextual modeling are presented. The chapter concludes with some thoughts about the nature of context quality as it may be applied to the confidence of processing and utilization of knowledge from contextual processing.

2.1 INTRODUCTION AND KNOWLEDGE DERIVATION FROM THE SNOW OF DATA

Today's world of information is critical to the control and operation of complex systems and humanity's welfare. It is utilized in every part of our lives. The development of computers and particularly the personal computer has led to an explosion of information that is accessible to a wide mass of people who have never historically had access to such information.

Coincidentally, there has also been an explosion in the amount of information that is being stored and used in decision-making processes. Originally, data were something that people and computers collected and then acted on in some fashion. As technology and models of information have developed, it has become increasingly evident that metaknowledge can be derived from the relationship among data and knowledge itself. Recently in research circles, scientists have become increasingly interested in the idea that information systems should change their activities based on the environmental information surrounding them. There are many terms for this, but for the purposes of this book we define this model to be *contextual-based processing*. As an example, if we are tracking a submarine as it is, we may merely know that there is a submarine underwater. However, if we have contextual information about its speed, depth, direction, and location, this becomes a context for deciding what potential intention or activities the crew on the submarine is involved in. To better understand the subtleties of context, we must first look at the nature of knowledge, how it is derived, and how it is interpreted to provide contextual metaknowledge.

The development of knowledge is a complicated process. It starts with the collection of data. Data comprise an abstract representation of reality that is based on how humans organize information. Data come in a variety of forms ranging from photographic to video, numeric, alpha, binary, geometric, and so on. So, the first step toward the development of knowledge is the collection and storage of data. Once data are collected, they are often not that useful until they are transformed into knowledge. This is because there can be intricate relationships among data that can lead to interpretations predicting the future, describing the present, and analyzing the past. The key is deriving knowledge, which requires the process of transformation and analysis to draw inferences that are

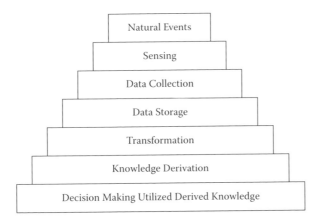

FIGURE 2.1 The process of turning data into knowledge.

the essence of knowledge. Thus, a definition of *knowledge* might be a paradigm that transforms raw data into the input for decision-making processes. Architecturally, the process of transformation looks like Figure 2.1.

In Figure 2.1 natural events are sensed, collected, and stored, and eventually knowledge is derived from the events that drive decision making. In this process, another model might be derived as shown in Figure 2.2, where data are affected in the transformational process by contextual data as they proceed to knowledge.

The process of conversion of data to knowledge often has temporal and spatial criticalities, especially in the context of natural disasters that define a context and the application of that context to processing. Sharing of information with people, governments, and entities at all the steps shown in Figure 2.1 is critical; and if this is not done in a coherent and contextual fashion, the results can be life threatening. To illustrate this fact, we examine one more manmade disaster to lead to the concept of dimension describing context.

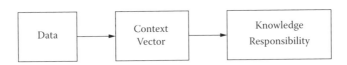

FIGURE 2.2 The architecture of knowledge development for context-based processing.

2.2 THE IMPORTANCE OF KNOWLEDGE IN MANMADE DISASTERS

There has been a recent shift in the computing paradigm away from the concepts of hardware processing data in a repetitive, structured, monotonic fashion. This evolution has become partly driven by some recent spectacular natural and manmade disasters. Among such disasters, we have seen 9/11, tsunamis in Asia, volcanic eruptions, and catastrophic nuclear accidents. In all of these events, there has been a static structure of information distribution, processing, and usage that has not kept up with the needs for knowledge and situational awareness based on the context in which the information was collected. Some initial work has been done in ubiquitous computation for such things as cell phones. However, there is a larger question of how to define a comprehensive global model of contextual processing that factors in the complexities of spatial extent and temporal events over a field of continuous and disparate data.

The question of contextually based processing has increasingly become how to interrelate information scattered across multiple organizations such as universities and governments to create a global integration of information. This is where the concept of information as a primary driver leads to the notion of information flowing to the appropriate resources, where it dictates the type of processing and analysis to be performed on it. Some of the issues are how to develop contexts that have security and integrity on information as it flows over the Internet, and how to manage its processing. In simpler terms, in an environment of hyperdistributed sharing, context-specific processing, political boundaries, and conflicting legal systems, how does one develop a comprehensive and secure model for doing advanced global context-based processing?

Previously, two disasters were discussed and the concept of contextual dimensions started to develop. These defining dimensions were *temporal*, *spatial*, *impact*, and *similarity*. To illustrate these dimensions in another context, the 9/11 disaster will be considered against some context-based properties (dimensions) that can affect how information is secured and processed.

2.2.1 September 11: World Trade Center

On September 11, 2001, the world witnessed the culmination of a plot that had been planned, coordinated, and developed over a long period of time to blow up the World Trade Center. In the information assurance (IA) lexicon, this attack had elements of what is often referred to as a *slow attack* with doorknob rattling. *Doorknob rattling* is trying different avenues to see which ones offer access to a situation.

Slow attacks are characterized by the probing and quiet penetration of a system where, upon a previously established time or triggering event, the attack initiates in a catastrophic, coordinated, and lethal fashion—in this case, the initiation of four directed aircraft in a highly synchronized attack modality. Doorknob rattling was the precursor of this attack and is, as mentioned above, a method of testing quietly to see what vulnerabilities exist in a system. In this case the terrorists flew on multiple flights, observing and testing aspects of the aviation security system and procedures.

The net result of this methodology was that at a little before 9 a.m., on September 11, people in New York City witnessed planes destroying the World Trade Center. The result of this well-coordinated attack led to a complete breakdown in situational awareness for people on the ground. Affected parties ranged from first responders and police to civilians trying to escape or standing around watching the spectacle. Each one of these groups had a need for varying levels of access to information that could be found scattered around in various pockets of organization-specific information repositories. For instance, civilians in the tower were slow to evacuate because they did not know visually what had happened despite images on television clearly showing planes and flames. Some stayed in the building waiting for an evacuation order as a result. This is an example of context. The televised images in conjunction with observation of the building shaking could have led to the conclusion that collapse was imminent and evacuation needed to be immediate. Had information about the attack flowed to the phones, portable digital assistants (PDAs), and computers of people inside the towers, they might have been more proactive in their response.

People on the ground were equally unaware of the dangers of building collapse due to a lack of contextual information to assign meaning to the event. They often merely stood around watching the collapse and blocking fire department access. A system where information flowed to where it was needed and provided knowledge based on context might have allowed first responders a few more minutes in the building to evacuate more people and saved onlookers from being killed by falling debris.

Partly because of the traditional methods of information processing, various aspects of information were often not present for the individuals and entities that needed the information. The result, sadly, was complete communication infrastructural collapse, loss of life, and general chaos.

The interesting aspect of the 9/11 disaster was that useful information was available in several different information systems scattered around the world. If it had been correlated and evaluated in the context of all information, this event might have been mitigated or avoided. For instance, the German police had information about the terrorist cell that the hijackers had come from. The terrorists were photographed on surveillance cameras in various places, the airlines had records of the many flights the hijackers made to evaluate aircraft security, and a flight school in Florida had observations of foreign students wanting flight lessons with no instruction on how to land or take off. If there had been a mechanism for all of this information to be collected and viewed in a comprehensive context, it could have suggested a concerning pattern of activity that could have been acted upon. In this case, more complete correlation and analysis were not possible because of traditional centralized processing. What could have helped is a model for correct and consistent information flow, aggregation, and information-controlled processing. Such a paradigm might be referred to as a *contextually driven processing model* that needs to incorporate the key characteristics of data, the dimensions of *impact*, *temporality*, *spatial impact*, and *similarity*. Borrowing from the examples in the previous chapters, examples of how these might be relatively classified for 9/11 are as follows:

- *Temporality*: Immediate.

- *Damage*: Extreme, civilization level, war.

- *Spatial impact*: Small, localized to an area downtown.

- *Similarity*: High—most sources of information were describing the same event.

The following section introduces the context model and key concepts of the model, examines data characteristics, and eventually develops the notions of contextual data, similarity analysis, and supercontexts.

2.3 CONTEXT MODELS AND THEIR APPLICATIONS

Recent years have witnessed rapid advances in the enabling technologies for pervasive computing. It is widely acknowledged that an important step in pervasive computing is context awareness. Entities in pervasive environments need to be aware of the environment so that they can adapt themselves to changing situations. With the advance of context-aware computing, there has been an increasing need for developing formal context models to facilitate context representation, context sharing, and semantic interoperability of heterogeneous systems. In this case, semantic interoperability means a standard method of communication about contexts that drive processing.

The concept of context has existed in computer science for many years, especially in the area of artificial intelligence. The goal of research in this area has been to link the environment in which a machine exists to how the machine may process information. An example typically given is that a cell phone will sense that its owner is in a meeting and send incoming calls to voicemail as a result. Application of this idea has been applied to robotics and to business process management [1].

Some preliminary work had been done in the mid-1990s. Schilit was one of the first researchers to coin the term *context awareness* [2, 3]. Dey extended the notion of a context with the idea that information could be used to characterize a situation and thus could be responded to [4]. In the recent past, more powerful models of contextual processing have been developed in which users are more involved [5]. Most current and previous research has still largely been focused on the development of models for sensing devices [6] and not on contexts for information processing.

In previous works, both informal and formal context models have been proposed. Informal context models are often based on proprietary representation schemes which have no facilities to improve shared understanding about context between different systems. Among systems with informal context models, the Context Toolkit represents context in the form of attribute–value tuples, and Cooltown proposed a Web-based model of context in which each object has a corresponding Web description. Formal context models commonly employ formal modeling approaches to manipulate context. Some work has been done using both entity relationship (ER) and Unified Modeling Language (UML) models for relational databases. However, none of the previous work has addressed a global contextually driven processing model for formal knowledge sharing, or has shown a quantitative evaluation for the feasibility of context reasoning, until just recently [7, 8, 9]. As mentioned previously, most previous work has centered on the application to hardware in somewhat limited environments such as the smart cell phone that shifts to voicemail in meetings.

The models developed and proposed in this book suggest that contextual processing can be based on the notion that computers can both sense and react based on their environment and more particularly on the scope and context of the data they are processing. Computers can have environmental information about what is occurring in the natural world from which they modify their operation using rules and heuristics. In this sense, they react intelligently to stimuli. This produces a more sophisticated paradigm for the derivation of knowledge from raw data. In this model, contexts are used to adapt interfaces, select relevant data for an application, drive the selection of information retrieval processes, locate services, and involve the user in the processing.

Additionally, contextual processing is concerned with sensing of the environment through acquisition of data. These data may be retrieved from databases, knowledge-driven processes, or physical sensors. Once information has been collected, it needs to be transformed into knowledge as previously discussed in order to control the processing and dissemination of information. The goal of this type of system is to adapt dynamically to situations where data might be highly heterogeneous, distributed, and widely describing events with different spatial and temporal events. Such a system will want to utilize many networks and computers simultaneously to derive information. In such a process, contextual processing, based on contextual data, is different from traditional processing models, where a single hierarchical entity may process only the data it has access to and control of. This old model is similar to the paradigm of desktop computers, where a single user works on a single computer perhaps unaware of computational resources and data that could be located on the next machine and could assist in his or her activities.

A key concept of contextually processing data is that the context model has dimensions to the information. As defined previously, they are as follows:

- *Temporality*: The time period in which the event unfolded from initiation to conclusion.

- *Damage*: The relative damage of the event in terms of both casualties and monetary loss.

- *Spatial impact*: The spatial extent, regionally, in which the event occurs.

- *Similarity*: The degree to which contextual data are similar and thus have the potential for aggregation.

2.4 DEFINING CONTEXTUAL PROCESSING

The previously discussed disasters could have been mitigated and responded to more effectively, and even perhaps thwarted, if an advanced model for contextual processing of information had existed. This type of processing is a new approach to information and knowledge derivation. As mentioned previously, an often-cited example of this is a cell phone that senses its environment and redirects calls when its owner is in a meeting. Contextual processing is similar to the simple cell phone example but considerably more elaborate in scope and computational power. In this model, temporal and geospatial components are considered for a widely distributed area over widely differing and disparate sources of data. Data collected for this new model would be transformed into knowledge via analysis partly as a function of similarity and aggregation.

Building on the concept of contextual dimension (similarity, impact, temporality, and spatial impact) is that in its simplest form, a context is formed from the collection of data that is transformed via analysis. At the core, contexts are stored in this model as feature vectors that can have mathematical properties defined upon them which can facilitate analysis. A *feature vector* describes an event of some sort that is typically described by the four dimensions of contexts discussed previously. These events are referred to as *event objects* and are discussed later on. A context's data are heterogeneous in the type of information they contain and can be denoted as

$$f = <a_1, a_2, a_3 \ldots a_n>$$

where $a_1, a_2, a_3 \ldots a_n$ are attributes of the context with data type x for an object whose context is being defined; and $f<>$ is a feature vector describing a context at time t_0.

The *attributes* of a context can be of any type of data relevant to the construction of the context and have characteristics utilized to derive contextual meaning. For example, they might be of the following characteristics:

- Numeric

- Textual

- Qualitative

- Quantitative

- Media such as audio, video, and imagery

Some services to which a context can be applied are as follows:

- Derivation of knowledge

- Creation of services

- Development of actions or responses

As such, a context also needs to have responses that correlate to the information contained in the context data. This can be denoted as

$$C_n = \{F_n, R_n\}$$

where F_n is the raw context data feature vector, and R_n comprises the contextual semantic processing rules for the given context that drive contextual processing.

Because a context's structure has a mathematical basis in vectors, the new context model has the property of union. This is an important property due to the fact that contexts can often be ambiguous over time, space, and impact dimensions. A context may, for example, be missing data in one or more of its attributes in its feature vector. It may be about a region of the earth that may be generated at a point in time t_0 where an adjacent area may be generating a conflicting context at the same moment in time. Because contexts are based on the dimensions of spatial and temporal properties in their generation, they can be continuously generated for an event or discretely generated, which can introduce problems (discussed in later sections of this chapter). As such, *contextual processing* needs the ability to combine contexts C for a given event object (an object producing contextual data) composed of feature vectors of attribute F and their rule sets R where there are *similarities* into supersets of contexts C'. The combination requires methods of reasoning about similarity (SIM), which is also discussed in subsequent sections of this chapter. The process of combination is characterized by Figure 2.3.

Of note is the cardinality of the relationship between C' and multiple contexts C. Additionally, the cardinality of the relationship of F with R may be any of the following:

$$1{:}1$$

$$n{:}m$$

$$x{:}y$$

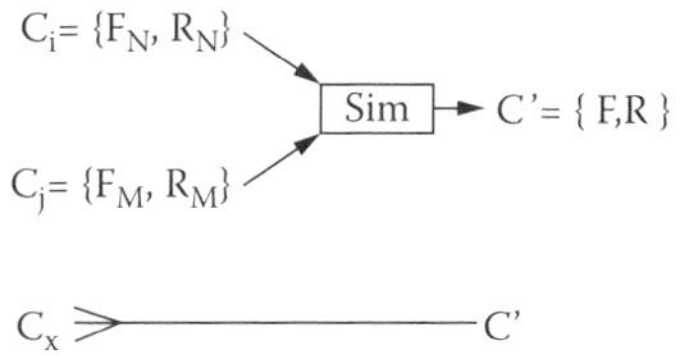

$$C_i = \{F_N, R_N\}$$
$$C_j = \{F_M, R_M\}$$
$$\boxed{\text{Sim}} \rightarrow C' = \{ F, R \}$$

$$C_x \!\!>\!\!\!-\!\!\!-\!\!\!-\!\!\!-\!\!\!-\!\!\!- C'$$

FIGURE 2.3   Aggregation of contexts based on a similarity function (**Sim**) into contextual supersets, where F is feature vectors of data and R is the contextual-processing rules for a given attribute in F.

where $n{:}m$ is a many-to-many relationship, and $x{:}y$ is a mapping of x attributes in the feature vector to y processing rules.

Some of the characteristic types of information that a context feature vector may contain can vary widely but may contain the following information for any given contextual event:

- Temporality of the event

- Spatial identity of the event

- Absolute reference to a point in the event space that allows ordering of context

- Historical perspective

- Subject classification of the event in a way that the concepts of subject in contexts can be compared analytically

- User profiles of key people involved in the event

- System profiles of processing systems affected by the event

- Interconnection sharing and communication among entities described by the event

- Ambiguous similarity reasoning capabilities, given conflicting information

- Automatic reasoning about the processing response

- Granularity, and macro-abstract representation of events where a lens can be utilized to look at an event from various points of magnification

- Multicontextual assimilation by similarity of N contexts

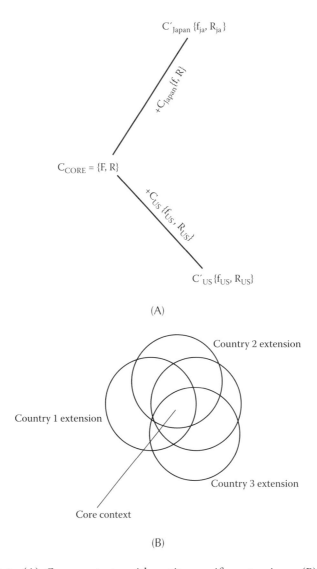

FIGURE 2.4 (A) Core contexts with entity-specific extensions. (B) Another method to visualize core global context.

Global contextual processing is a term that can span multiple agencies and governmental entities. This leads to another core idea in the model: the notion that there cannot be the concept of a *universal context* because of the widely varying security, legal, and geopolitical constraints on global information sharing. With this in mind, it is probably possible to develop a core context that can be shared on a global scale with entity-specific extensions that address localized information policies and requirements, as shown in Figure 2.4.

To better understand complex and key issues in a context model, it is important to have a fine understanding of the characteristics of data found in such a model and how that might affect contextual processing. The next section examines in detail the nature of data that might be utilized in a contextual-processing model.

2.5 THE PROPERTIES OF CONTEXTUAL DATA

Knowledge is derived from data, and data comprise a representation or abstraction of an observable event. To understand more ideally what this means, we examine in this section the types of data that can be collected for a given context and the attributes of such information. In this discussion, we will discover that data are often *crisp* in that they represent a single value and other types of contextual information can be *noncrisp*, representing the vagaries of how humans describe situations in the real world.

There are many fundamentally defined types of data. If one asks computer scientists about the types of data representation they utilize, they might reply *binary*, *numeric*, and *alpha*. These are the fundamental types of information representation found in a computer utilizing the base 2 number system (binary).

Of course, the fundamental data types are aggregated to create numeric systems of integers and real numbers, which are then aggregated to create more complex types of data such as images and audio. This relationship can be found in Figure 2.5, where the lower basic

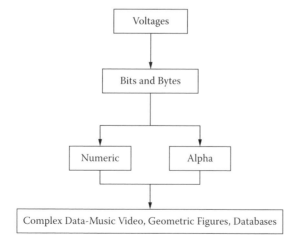

FIGURE 2.5 Aggregation of fundamental data representations into complex structures.

TABLE 2.1 Operations Defined for
Quantitative and Qualitative Data

	Operators
Nominal	$= \, !=$
Ordinal	$> \, <$
Interval	$+ \, -$
Ratio	$+ - / *$ ratio

data types are combined into the specification of more complex types of information.

The fundamental data types can be characterized by the *properties* of the information they represent. The first of these has to do with the *type of information* being described; this is referred to as qualitative and quantitative.

Qualitative data are categorical in nature and are broken down into the following categories:

- *Nominal*: Data that cannot be ordered, Examples are zip codes, IDs, and genders.

- *Ordinal*: Data that can be ordered. Examples are grades and street numbers.

Quantitative data are different from qualitative data in that their scheme is based on relations among numeric data. Qualitative data also have two subcategories:

- *Interval*: A scheme where ratios are meaningless but differences are meaningful. Examples of this type of data are dates and temperature information.

- *Ratio*: A scheme where differences and ratios are meaningful. Examples of these are money and weight.

Various mathematical operations are defined upon these types of data. They are shown in Table 2.1.

2.6 CHARACTERISTICS OF DATA

In addition to the properties and types of data, there are characteristics that can be used to describe how data are interpreted. For example, in integer number systems, integers do not have a numeric value between two adjacent numbers. There is no concept of the value 1.5 in an integer-based number system. People use such a number system because its storage representation is a lot smaller than that of other number systems and the binary nature of the information is handy for representation of the presence or lack of presence of information such as a voltage. It is also faster to complete mathematical computations on integer data. This type of data is analogous to discrete data because intermediate values on an interval are not defined.

Discrete data occur in fixed regular intervals and cannot be subdivided. In a sense, this type of data is atomic. The numbers 1, 2, 3, 4, and so on are examples of discrete data. The assumption that this type of data makes is that it cannot be subdivided and values don't exist or can't be measured in the interval between the discrete values of the data. As mentioned earlier, discrete data are convenient for storage and calculation. Sometimes discrete data systems are used because a sensor or technology does not have fine enough resolution to sense portions of real-world data values between the discrete values.

In contrast, continuous data can have intervals between values endlessly subdivided into smaller and smaller values. This process can be extended to the limits of infinity. For example, if one is asked to find the number halfway between 1 and 2, in a continuous system the value 1.5 would be the answer. A sensitive instrument can, within certain operational limits, sense continuousness to a given limit. If one is asked to find the number halfway between 1.5 and 1.6, a value of 1.55 can be presented in a continuous system. This process has the interesting characteristic that the process of finding numbers halfway between two values can continue to infinity, producing smaller and smaller numbers. In other words, there is no limit to numbers that can be represented in a continuous system. This type of data corresponds with the operations found in the real number system R. Because the data can be subdivided infinitely, most sensors that collect such information have operational limits below which physical-world data cannot be sensed. This leads to another characteristic property of data, that of interpolation.

Interpolation of data (see Figure 2.6) is a method of estimating approximately what a value might be when it cannot be actually sensed or observed.

$$(Continuous) Estimated\ Value = ((m-n) * .5) + n$$

FIGURE 2.6   Interpolated estimation of a value for continuous data.

There are a variety of methods for doing interpolation. One of these is to look at the distance between two numbers as a ratio. For example, given numbers *n*, *m*, if asked for the value halfway between these numbers (.5), a ratio for estimating the number can be derived from an equation like

$$(Continuous)Estimated\ Value = ((m-n)^*.5)+n$$

Because of round-off error commonly found in computer-based calculation, the exact value of an interpolated calculation can never be known exactly or with certainty. Another characteristic of data is the degree of dissimilarity they can have. This is the apples versus oranges comparison problem. In this characterization of data, two categories of data can be found: *heterogeneous* and *homogeneous*. These classifications apply to all the types of data mentioned previously.

Because data can be sensed and collected about many different topics, it is possible to get various types of data about a theme or a subject area. In GIS, these data are referred to as *thematic data* and can often be found when describing geospatial and temporal phenomena. This type of relation characterizes data based on their relationship and similarity or dissimilarity. For example, if one represents the extent of a natural disaster in a spatial data system, the system will contain lines and points representing map features; it may contain polygons that overlay the map where the inside of each polygon is a discrete value representing, for instance, average wind speed and another overlapping polygon represents a discrete value of housing damage in dollars.

In such a thematic system, we start moving away from merely having data to *interpreting* the data, and *transforming* it into knowledge. For example, in Figure 2.7, the wind speed of *x* for a given region correlates

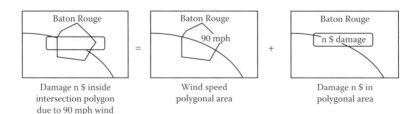

FIGURE 2.7 Thematic polygon overlays; intersection and interpolation are used to deduce the average dollar costs of damage for a region based on wind speed.

to damage values of y. Therefore, we can infer knowledge that where we see wind speeds of a given value (90 mph), the damage to housing will on average be $y(100)$ dollars.

Heterogeneous data, however, can be complicated to compare. If one creates a scenario where we are mixing ordinal, nominal, interval, and ratio data for two different object entities, it becomes difficult to make a determination of how similar or dissimilar the entities really are. This is like trying to make a judgment about how closely related an apple is to an orange or comparing a picture to a number. Fortunately, there are methods found in data mining that help with these sorts of comparisons.

Homogeneous data are often easier to compare than heterogeneous data because the units that define the data are the same. In other words, it is simple to compare two temperatures and evaluate their relationships as, for instance, greater than, less than, or equal. However, it is another question to determine if a temperature has a correlation to wind speed or wave height. Homogeneous data provide simpler information that is easier to compare, but it is harder to get a comprehensive view of information because such data only describe a certain type of information. Methods that allow disparate standardization and thus comparison of heterogeneous data provide better knowledge granularity.

The above are a few characteristics of data that need to be considered in the creation of a model for contextual processing. However, data also can have complex characterizations based on temporal and spatial properties. This is the subject of the next section.

2.7 SEMANTICS AND SYNTACTICAL PROCESSING MODELS FOR CONTEXTUAL PROCESSING

In addition to the complexities of data classifications and categorizations, contextual data can often have a complex geospatial and temporal aspect to it that can affect contextual processing. This property is key in the definition of two of the four dimensions defining contextual data and its processing. Contextual data often represent complex events. Events can have multiple temporal and spatial characterizations that can drive the derivation of a semantic model that can be utilized for contextual processing. These characterizations can have a relation with the *similarity* dimension and drive the immediacy of the processing and dissemination of knowledge. Some of the characterizations describe spatial properties, and some describe temporal dimensions of objects. Some characterizations of spatial properties for objects and events producing contextual information are as follows:

- *Singular*: An event that happens at a point in time, at a singular location.

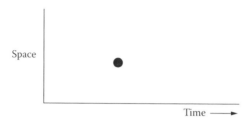

- *Regional*: A single event that occurs within a given region at random locations.

- *Multipoint regional*: A multipoint event occurring within a region.

- *Multipoint singular*: Events that occur at a single point in time but with multiple geographic locations with no particular regional association.

- *Affection*: The spatial extent to which events affect their surroundings.

- *Directionality*: A descriptor of the way a spatial event is spread in time and space. This may be linear like a rocket or polygonal as in the case of a spreading tsunami wave.

Contextual data produced from objects (e.g., sensors or observations) also have patterns in the nature of information creation over time. Some characterizations of temporal behavior of contextual objects follow. The arrows are meant to symbolize the amount of information content coming from a given event object at a point in time. The bases of the arrows are positioned on the timeline at the moment that the event object generates information.

- *Episodic*: Events that occur in bursts for given fixed or unfixed lengths of time and have some information content, or a burst of contextual information.

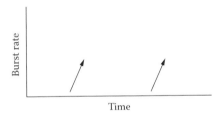

- *Regular*: As suggested, these events occur at regular intervals giving off bursts of information.

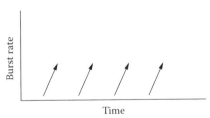

- *Irregular*: The time period between events of this type is always changing.

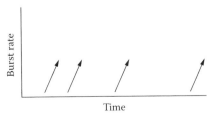

- *Slow duration*: A series of event(s) that occupies a long duration, for example the eruption of a volcano, where the event has a beginning and end and bursts information continuously during the interval.

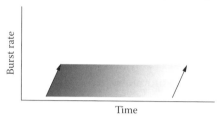

- *Short duration*: Relative to slow-duration events, these occur in a short duration, for example an earthquake that last 30 seconds versus a volcanic eruption.

- *Undetermined*: This is the null condition where the duration of an event object is not known or cannot be determined.

- *Fixed length*: Another method of specification similar to a bounded event, but by default a fixed-length object must be bounded. However, *bounding* refers to a fixed point in time, whereas *fixed length* refers to a length of the time period of an event that is known.

- *Unfixed length*: Essentially, this is an unknown length of time that an object may burst information. It may be infinite, and/or the length of periods between bursts may change.

- *Bounded*: A bounded event occurs between two points in time that are known. In the following, the burst event object occurs between bounds zero and ten on the time scale.

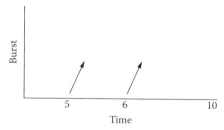

- *Unbounded—continuous*: This type of bursting event object can be analogous to an undetermined characteristic or is infinite in length.

- *Forward limited*: These are burst events for which the origin in time is unknown and the point of cessation in bursting is known in the future.

- *Backward limited*: This is an object that has a point in time of burst origin but no known point in the future at which bursting halts.

- *Immediacy*: A descriptor of how immediately adjacent an event might be to a point in time. It can be defined with impact to produce characterizations such as *urgent* or *catastrophic*.

- *Repetitive*: This is similar to the concept of a regular event except that repetitive temporal events can product several bursts for a short time, then stop and restart at some point in the future. A repetitive event is, therefore, a type of regular event.

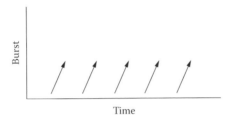

The previously defined characteristics (dimensions) of a contextual analysis of the 9/11 disaster were as follows:

- *Temporality*: The time period over which the event unfolded from initiation to conclusion.

- *Impact*: The relative damage of the event in terms of both casualties and monetary loss.

- *Spatial impact*: The spatial extent, regionally, over which the event occurs.

- *Similarity*: The degree to which sets of collected contextual information are related to a given theme.

These dimensions can be described and characterized by the properties of data discussed previously to provide a rich semantic meaning and thus a context for processing into knowledge. As such, they are components of a contextually based processing model.

The above discussion can be utilized and extended into a semantic model that explains or identifies how the processing rules associated with a context are applied to determine processing and security. Based on the previous notions, semantic characterization classes of information can be derived which can then be turned into a processing grammar that can drive the context transformation engine during the process of the transformation of data to knowledge. The following semantic model contains elements from the discussion above that can affect context-based processing. They are organized into metacategories as follows:

> Event Class <abstract, natural>
> Event Type <spatial, temporal>
> Periodicity <regular, irregular>
> Period <slow, short, medium, long, undeterminable, infinite, zero>
> Affection <regional, point, global, polynucleated, n point>
> Activity <irregular, repetitive, episodic, continuous, cyclic, acyclic>
> Immediacy <catastrophic, minimal, urgent, undetermined>
> Spatiality <point, bounded, unbounded>
> Dimensionality <1, 2, 3, n>
> Bounding <Fixed Interval, Bounded, Unbounded, Backward Limited, Forward Limited, Continuous>
> Directionality <Linear, Point, Polygonal>

FIGURE 2.8   Modeling the semantic categories of context-based meta classifiers.

The above metacharacterizations capture the notion of similarity (or dissimilarity) by spatial and temporal properties of contextual data. Of note, the previous characterizations of data can also apply to every element in a context feature vector ($f<>$), and thus the power of the model can be expanded considerably. A semantic syntax based on Figure 2.8 can be derived that can then create processing sentences driving contextual processing and response. It has the following set of production rules of the following form:

> R1: <event class>, <event type>, <R2>
> R2: (<periodicity> <period>) <R3>
> R3: (<affection> <activity>) <spatiality> <directionality> <bounding> <R4>
> R4: <dimensionality> <immediacy>

FIGURE 2.9   Syntax for application of semantic modeling affecting contextual processing.

The above grammar is based on metaclassifications reflecting the semantic similarities of data in a context, and thus it also can lend itself to the analysis of thematic similarity discussed in subsequent sections. The above grammar and syntax can be used to develop sentences that affect and drive the operation of contextually based processing. For instance, a tsunami in the Indian Ocean might generate a *contextual-processing sentence* such as

> *R1 = natural, spatial-temporal, irregular-slow, regional episodic catastrophic unbounded 3D linear.*

The application of this model and its production rules can then be mapped to resource actions affecting such things as priorities on use of information, computational resources, dissemination and notification, and so on. For instance, the above production *R1* may produce the following response rules affecting the processing and/or collection of contextual information:

> *natural spatial temporal => {notify and activate associated computational resources}*
>
> *irregular-slow – tsunami => {notify hotels, activate alert system sensors, satellite tracking data computers activated, and high-priority collection mode}*
>
> *catastrophic => {context processing elevated to kernel mode}*
>
> *unbounded => {distributed notification and transfer of context processing to surrounding countries}*

Note that the left-hand side of the above response rules is open architected such that an entity (government, military, etc.) might place the rules or specifics of how to respond to, for instance, a *natural spatial temporal* sentence component. There can be, and are, many other classifiers of contextual events that could affect the processing of a context's data and thus affect its transformation from data into knowledge or application. However, new classifiers should fit into a member of the semantic model proposed above. Even more importantly, the selection of the transformation method to knowledge will be predicated on the characteristics of the event's context, as it is sampled over time for various geographic extents.

Another factor in a contextual-processing model to consider is that events occurring over time can diverge from root events in the types and themes of data being collected. Thus, contextual data and their processing can diverge but still have relationships to other contextual data at previous locations in time. Divergence can create *N* levels of parallel contextual threads that are similar or not similar, or for which the classification of the

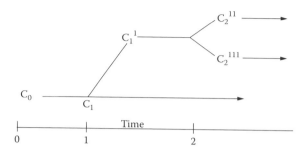

FIGURE 2.10   Divergence of data events flowing through time and space.

context is not known. Additionally complicating the matter is that these *contextual threads* could have many root nodes. The relationship among related contextual threads is also important to consider in the derivation of context-based knowledge. Understanding of the relationships or lack of relationships among threads can drive inference and aggregation modeling of contextual processing to produce new, unseen conclusions and processing responses, and thus new knowledge.

As an example of this, consider Figure 2.10 and the notion that at time t_0, an earthquake occurs in the Indian Ocean. Information from multiple sensors about the earthquake might be magnitude, directions, and so on. This earthquake may lead to several related contextual data streams following in time and spatial extent after the initiation event. For example, subsequent earthquakes could be generated, landslides could occur, and tsunamis may flood coastlines as shown at time t_1 and t_2. All of the above could generate streams of data whose relationship needs to be understood and may be included in a *supercontext* of processing rules and data that is much more powerful than an individual thread's information. What is of particular note is that the flow of context event threads in time and space has interesting properties similar to some of those previously suggested, such as being the following:

- *Forward limited*: A temporal event that has no clearly discernible beginning but does have a terminal end point in time

- *Backward limited*: A temporal event that has an initial point of inception but no foreseeable end in the future

- *Continuous*: A temporal event that has no clear beginning and no clear end

These can become additional metacharacterizations useful for the notion of contextual-processing threads that diverge. These properties also need to be considered in the derivation of knowledge within the contextual model. Key to understanding is that there may or may not be a clear and definable point in time when the event comes into existence or ceases to continue. So, despite knowing that the universe probably had a beginning and will probably have an end, it can at best by the above definitions be classified as a continuous threaded event because there is no termination or initiation that is exactly known in time. Thus, data describing such threaded contextual events should reflect the properties mentioned above. Of note is that these properties typically can only be applied to the temporal dimension of contextual processing of natural disasters; they will not easily be applied to data describing the spatial extent of a contextual event, which will typically be bounded in some fashion.

2.8 STORAGE MODELS THAT PRESERVE SPATIAL AND TEMPORAL RELATIONSHIPS AMONG CONTEXTS

The ways that contextual data are organized after their collection comprise another key consideration in the derivation of a contextual model. Due to the previously mentioned spatial semantics, the relationships must be captured and may not be crisp in meaning. The key to storage organization is that it must also maintain the geospatial and temporal relationships among widely disparate data types found in a context feature vector.

As methods for classification and derivation of knowledge from a context are developed, contextual characteristic operators can be applied to spatial and temporal context-based data to reason about the information. Retrieval of a context's information can then be expressed in spatial and temporal queries of information. In order to have such a query, data should first be organized into a model structure that preserves temporal and spatial relationships in such a way that contextual retrieval and analysis are facilitated.

A model that has the possibility of maintaining such relationships is to think of contexts and contextual data being organized into cubes where each layer in the cube has a thickness or no thickness at all. Thickness in this case could have semantic meaning such as the degree of contextual importance of a data layer in knowledge derivation. Additionally, layers could be thematic in nature. In this model, cubes can also be ordered and thus express the concept of similarity by order. This would function as a contextual similarity based on distance from a particular layer in the cube.

2.9 DERIVING KNOWLEDGE FROM COLLECTED AND STORED CONTEXTUAL INFORMATION

A semantic grammar has previously been proposed as a framework to express contextual processing and thus reasoning about the data and their importance, dissemination, and so on. This section looks at derivation of metaknowledge from contextual data that might further augment the derivation of processing knowledge.

The process of deriving knowledge from information can be accomplished in a variety of ways; however, the process is generally the same. First, one has to decide what data to collect. This is a critical step that must be clearly defined before anything else is done. The definition of what is to be collected is derived from deciding very specifically what question about a theme one will attempt to answer. The next step in this is the actual collection of data via sensors. The bits and bytes collected in our model thus far represent the natural world as an abstraction of reality. Because data collection can be prone to errors in sampling, it is often a good idea to collect several sets of data that are representative of the question that one is trying to answer. The next step in the process is cleaning the data. This is necessary because sensing and generating data can create anomalies in the data. These are often referred to as *outliers*. Outliers are anomalous values that deviate from the general collection of data which can show up in contextual information due to mis-sensed information. Figure 2.11 is a visual example of how outliers may be plotted. Outliers need to be removed from data because if allowed to enter the analysis process, they can heavily skew the result of the analysis and thus provide erroneous information.

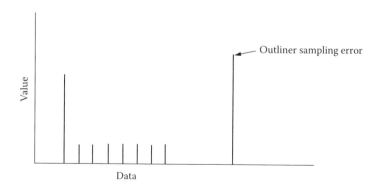

FIGURE 2.11   Example of outliers in data (chart with points).

The next step might be reducing the dimensionality of the data. *Dimensionality* is simply the number of attributes we might have, for instance in context E. For example, if we have a data event entity E, it may consist of a number of attributes of data creating an N tuple as shown here:

$$E = <a_1, a_2, a_3, \ldots a_n>$$

This is the idea presented earlier for the collection and storage of disparate data types about a given event in time. E will have several characteristics related to the earlier present dimensions of contexts. Specifically:

1. It will tend to change in content over time.

2. Some attributes may conflict in information content and semantics over time.

3. The data in the vector will have a theme (e.g., tsunami, 9/11, and so on, and since data).

The collection of data sampled over time for many event objects E can be aggregated into a set S, then becomes a collection of entity attributes and can be denoted as

$$S = \sum Ei = <a_1, a_2, a_3, \ldots a_n>$$

where S becomes a set of contexts representative of what can be referred to as the *universe of discourse*. Aggregation of E into S can have problems of its own which are discussed later. Loosely speaking, the universe of discourse is all information-producing entities that can be considered for a given thematic domain. In such a collection, one may not want to attempt to derive contextual knowledge by considering the use of all entities and all their attributes because it will probably be too computationally expensive, especially if the data are streaming over time. Thus, one borrows from data-mining concepts, selects a set of the sampled entities, and determines which sets of attributes provide the best discriminatory power for building a contextual classification model. This is referred to as *reducing the dimensionality* of the problem. Reduction of dimensionality can introduce the issues of false positives and false negatives in interpretation and reasoning about a context may semantically represent. This is because the data being analyzed are probably not all of the information found in the thematic universe of discourse.

More specifically, a *false positive* refers to a case where a condition appears to exist but does not actually exist. In logic, this can be expressed as an implication such as

$$e \rightarrow c$$

In the above, if event *e* is true, then this makes condition *c* true. For example, if in a test for cancer the event *e* comes back as true, then the condition *c* is that a patient has cancer. A false positive can be expressed in the form of

$$e \neg \rightarrow c$$

where the existence of an event appears to be true, for example a test for cancer says that a patient has cancer when in fact he or she does not have cancer.

A *false negative* is the opposite of a false positive. It is a condition where a test event *e* indicates that a condition *c* does not exist when in fact the condition does exist. In this case, a test for cancer would return the statement that a patient does not have cancer, when in fact he or she does have cancer. The same argument may be made about the detection of a tsunami.

Which one of these errors is most and least desirable can depend on *the impact dimension of a context* and implications of misclassification. In a cancer test, we want to error on the side of safety; therefore, we want to minimize false negatives as much as possible because otherwise people who have cancer will not get treated. A false positive might be better under the theory that sometimes it's better to be safe than sorry.

After reduction of dimensionality is done, an analysis method tailored to the dimensionality of contexts should be selected. In fact, it is often a good idea to have in mind what type of analysis to use when defining the problem to be solved because that will affect the type of contextual data being collected, the methods of collection, and understanding similarity in semantically diverging but thematically related contextual information. Overall, the processing of developing contextual knowledge from contextual raw data can be summarized as shown in Figure 2.12.

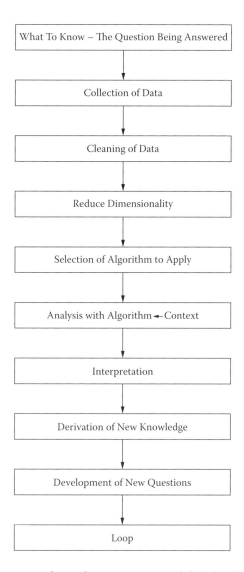

FIGURE 2.12   A process of transforming contextual data (E, S) into knowledge.

2.10 SIMILARITIES AMONG DATA OBJECTS

In order to reason about contexts and derive knowledge from them, there is a need to be able to find the similarities among contexts and thus generate new types of action responses to contexts. In addition, similarity analysis allows one to mitigate the effects of missing, ambiguous, or conflicting information for a given attribute (e.g., a_1 in two different feature vectors describing a given event such as a tsunami).

To illustrate new rule generation via similarity analysis, consider that given two contexts C_1 and C_2 with processing action rules R, we might be able to find the degree of similarities and then construct aggregate rule sets for the purposes of controlling processing. This approach can be summarized mathematically as follows:

Given C_1, C_2,

$$\text{similarity } (C_1) \approx \text{similarity } (C_2),$$

for feature vectors

$$C_1 = <a_1, a_2, a_3, a_4 \ldots>: R_1 \quad \text{and} \quad C_2 = <a_1, a_2, a_3, a_4 \ldots>: R_2,$$

where for all $Ci.aj$ homogeneous $(C_1.ai \; C_2.ai) \approx$ true,
  then

$$R' = R_1 + R_2 + R_n.$$

where R_1, R_2 are context-based processing rules and thus a type of processing knowledge for a given supercontext's data (S). If the two contexts are found to be similar, they can also be aggregated into C in a contextual model given earlier as $S_c = (C, R, S)$. As mentioned before, aggregation into C by similarity of context feature vectors can mitigate imperfect or conflicting information about an event.

Methods of calculation of similarity are sometimes referred to as *distance measures*, and are relatively simple to calculate for two different contexts. However, it is more powerful to consider the similarity of thematically related *sets of contexts* where large processing knowledge rule bases can be derived from component contexts to control processing. Given the following context $S_1 \ldots S_n$, it could be useful to have noncrisp methods employed to reason about thematic similarity.

2.11 REASONING METHODS FOR SIMILARITY ANALYSIS OF CONTEXTS

The development of powerful contextual models needs to also consider the fact that context feature vectors can be ambiguous, conflicting, missing information, or unable to be classified. This can be due to a number of factors found in the collection of context feature vector data, including the following:

- Continuous or discrete sampling
- Multipoint spatial sampling
- Multipoint temporal sampling
- The characteristic classifiers of data
- Divergent contextual threads
- Highly heterogeneous disparate data types found in a sample vector
- Geopolitical and legal aspects on what can be included in a context

Therefore, it can be difficult at best to say a given event's data context C is the correct information for an event and therefore should control processing. Additionally, recall that contexts are associated with processing rules which control the application of the context. Therefore, it is useful to aggregate contexts into *sets of supercontexts* based on the degree of similarities found in the set members. This can mitigate the effects of ambiguous or missing data. This section will evaluate some useful methods for aggregation within a theme such as data about a tsunami wave. Once aggregated into a set, it is possible to generate a supercontext vector that is representative of the context and a superrule set that then applies new contexts as they arrive.

Before contextual-processing knowledge can be derived, contexts need to be analyzed for similarity with other existing contexts, and aggregated with other contexts into data supersets of related contextual information. This process is necessary to reduce redundancy and allow for the combinatorial power derived from combining similar contexts and thus the contextual-processing rules associated with a given context. This process is referred to as the *creation of supercontexts by aggregation*. The process of doing the analysis of similarity of a new context against existing ones needs to determine if a new context's data can be determined to be the following:

- Part of a context
- Similar to a context and thus the same context

- Similar to a context but not the same context

- Dissimilar to a context

- Unclassifiable and therefore undeterminable

Multiple contexts may contain a wide variety of heterogeneous data, as discussed earlier. Thus, from elementary data, it is important to understand how to analyze for similarity and thus be able to aggregate based on a measure of similarity.

Multiple methods of similarity analysis exist. Often they classify a new database based on mathematical models and seem to come in two different categories. Some of the methods seek to build models that can classify new data based on inequalities and heuristic reasoning, and other methods look for the degrees of similarity by deviation from a mathematical moment.

Some *degree of similarity* methods that can be applied to contextual aggregation include the following:

- Fuzzy set membership reasoning

- Means, averages, and floors and ceilings

- Standard deviation

- Probabilistic reasoning

- Regressions

Some ideas from set theory can be directly applied to aggregation via the notion of crispness. Set theory is one of the most straightforward techniques for building contexts. In this case, cut levels of similarity could be equated to crispness in the determination of membership or nonmembership of an unclassified context in a supercontext. In traditional set theory, given a set S,

$$S = \{1, 2, 3, \ldots\},$$

the question of the number 1 being in the set is true. This can be denoted by a function such as the following:

$$member(1, S) = \{1, 0\}$$

which will evaluate as true (1) or false (0), and thus reasoning about membership can determine similarity, which drives the reasoning to aggregation of an unknown context into a supercontext.

2.11.1 Statistical Methods: Means, Averages, Ceilings, and Floors

Statistical measures attempt to characterize aggregated data and thus can be used to reason about similarity. A mean is simply found by dividing the sum of the elements in a set by the number of elements to produce a statistical moment that represents the aggregate value of all elements in a set. Once a mean is calculated, it is simple to compare the means of new contextual data and thus reason that the data are similar or dissimilar and thus drive the processing of aggregation of the data based on the result.

Ceilings comment that a value cannot be greater than some arbitrary value. Floors comment that a value must be greater than some value. Often equality operators of $<$, $>$, $\leq$, and $\geq$ are used in the construction of such functions. Ceilings and floors can often work in conjunction with each other to reason that contextual data are similar or not similar. Complex statements utilizing ceilings and floors are often created using AND ($\wedge$) or OR (v) operators into rule sets. Here are a few examples given value v and these operators:

If ceil[v, ceilvalue] = true ....v is less than the ceiling value.

If floor[v, ceilvalue] = true .... v is greater than the ceiling value.

If ceil[v, ceilvalue] = true $\wedge$ floor[v, ceilvalue] = true ....v is in a range between ceiling and floor.

If ceil[v, ceilvalue] = false $\wedge$ floor[v, ceilvalue] = false ....v is outside a range between ceiling and floor.

2.11.2 Fuzzy Sets

Fuzzy sets are derived from fuzzy logic and describe membership in a set that cannot be enumerated as true or false. Fuzzy sets are not crisp in nature, and thus membership classification determines the degree of possibility that an unclassified context's data may belong to a supercontext. Values for membership are continuous over a range between 0 and 1, where 0 means *not a member of the set* and 1 means *clearly a member of*

$$FnS() = \begin{cases} 1 \mid \text{if } a \in \text{domain}(A) \\ 0 \mid \text{if } a \notin \text{domain}(A) \\ [0,1] \mid \text{if } a \text{ is a partial member of domain}(A) \end{cases}$$

FIGURE 2.13   A characteristic function example.

the set. Because membership values are continuous data, they cannot ever be fully specified in traditional set notation and thus must be described in a different fashion. The notation for description of a fuzzy set is that of a characteristic function and has the following form:

$$FnS() = \begin{cases} 1 \mid \text{if } a \in \text{domain}(A) \\ 0 \mid \text{if } a \notin \text{domain}(A) \\ [0,1] \mid \text{if } a \text{ is a partial member of domain}(A) \end{cases}$$

where $u(s)$ is a characteristic function returning a value between 0 and 1. See also Figure 2.13.

The challenge of fuzzy set theory and application is that the interpretation of a value such as .89 membership in a set is very subjective, and thus it is up to the user to assign a semantic meaning to it. Some might say that a value of .89 means that there is a high possibility that these new contextual data belong to a given supercontext. Others might say that there is a good possibility of membership. The question, then, becomes "How does one compare *high* and *good*?" In other words, the question of similarity becomes subject to interpretation.

2.11.3 Standard Deviation

Standard deviation is another way that contextual data may be analyzed for similarity and thus membership and aggregation into a supercontext. In simple terms, standard deviation says that a percentage (e.g., 90%) of data will be ± a value, the standard deviation, from the average of a set.

What is key to the analysis is to have standard deviations that are a small distance from a given average, as shown in Figure 2.14. A tight curve plot of all sampled data means that information is more representative of all actual data and therefore better suited for gaining understanding and reasoning about the similarities in contextual data. The plot shown in Figure 2.15 is not as good for reasoning about similarity between contextual data and thus their possibility of being aggregated.

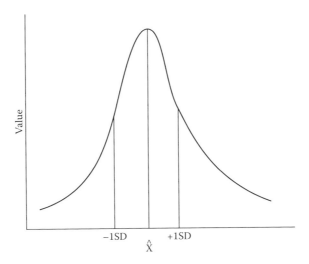

FIGURE 2.14 Graph of data that are closely related, with a small standard deviation.

In the application of standard deviation analysis, the method could be applied in a variety of fashions including calculation of characteristic means and deviations for all attributes in a contextual vector $<a_1, a_2, a_n>$ and coming up with an average standard deviation that could be utilized as a decision factor for the decision to aggregate.

2.11.4 Probabilistic Reasoning

Probability theory was developed by the French mathematicians Blaise Pascal and Pierre de Fermat in the seventeenth century, initially as a way

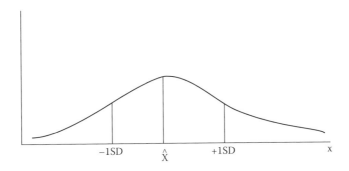

FIGURE 2.15 Graph of data that do not have good relationships, with a larger deviation.

to calculate the chances of a card being dealt. It has grown since then to apply to many areas of science and engineering, and its application can be applied to analysis of probable similarities among context feature vectors and thus candidacy for aggregation. The chain of reasoning might be something along the lines of the following: the context vectors f_1 and f_2 are within a high degree of probability of being the same and therefore should be aggregated into the supercontext vector S.

Probabilistic reasoning is similar to inductive reasoning, which states that one can predict the nature of the universe from observed events. For example, if one touches ice and it feels cold 90% of the time, one can induce that 90% of all ice will feel cold. Thus, this type of reasoning infers the likelihood of change, and an event will occur based on the odds of occurrence. For example, one might reason that most feature vectors from the Indian Ocean at a given time period are about a tsunami thematic object; therefore, other vectors occurring around that time period are *probably* also about the tsunami and should be aggregated.

The rest of this subsection provides a brief tutorial and some background on the subject. The probability that n particular events will happen out of a total of m possible events is n/m. A certainty has a probability of 1, and an impossibility has a probability of 0. In this way, probabilistic reasoning is similar to that of fuzzy set theory but differs in that fuzzy set theory comments on the *possibility* of a contextual feature vector belonging to a set, whereas probability makes a comment about the *probability* of belonging to a set.

Probability can be defined as "number of successful events/total possible number of events." As an example, when tossing a coin, the chance that it will land *heads* is the same as the chance that it will land *tails*, that is, one to one or mathematically even. Using the above equation, the probability is calculated as 1/2 or as a value of 0.5. An estimate for probability can be achieved by doing an experiment to calculate the frequency with which an event occurs. The probability of any given number coming up on the roll of a die is 6 to 1. Thus the probability of a particular number appearing on any given toss is 1/6 or 0.1666. If two dice are rolled, there are $6 \times 6 = 36$ different possible combinations. The probability of two of the same number occurring (e.g., 1,1 or 2,2) at the same time is six out of thirty-six possible combinations.

Events are considered *independent* if their occurrences don't affect the chances of other events occurring. The rolling of the two dice is an independent event, because the rolling of the first die does not affect the chances of any given number on the second die appearing. Opposite of independence is the concept of being *mutually exclusive*, which simply means that an event occurring prevents another event from happening. For example, tossing a coin is a mutually exclusive event because only a head or tail can occur but not both at the same time. Another property of probabilistic reasoning is that the probabilities of mutually exclusive events always add up to 1. For example, if one has a bag containing three marbles, each of them a different color, the probability of selecting each color would be 1/3. Such properties are also useful for reasoning about aggregation of contextual event feature vectors. For instance, the probability of the ocean being in a normal condition when much data about a tsunami have been collected at a moment in time could then be mutually exclusive and reasoned to be erroneous and thus not a candidate for aggregation.

An event is conditional when the outcome of a previous event affects the outcome of a subsequent event. For example, if a marble is chosen at random from a bag of four green marbles and five purple marbles, and not replaced, the probability of selecting two green marbles is that given in Figure 2.16.

Probabilistic reasoning could be utilized with contexts in a variety of ways. For instance, if a series of attributes describing a context feature vector exists and they have been aggregated, then this property might give an optimum suggestion of the best set of contextual-processing response rules, R, that should be selected to drive processing. In constructing a context, probabilistic reasoning might also be utilized to argue for which context a set of observed contextual feature vectors should best be aggregated into. Another example of application of this property might be to reason that if a large number of context feature vectors arrive about a tsunami after a set of vectors about an exploding volcano, then there is a conditional probabilistic link among the thematic events.

Having looked at some mathematically based methods that could be utilized for analysis of similarity in feature vectors and thus aggregation, we can consider methods that are more oriented to heuristically based classification and might be useful in similarity analysis. Some potential methods, but not an exhaustive list, might be

- Clustering
- Support vector machines
- Decision trees
- Bayesian techniques

$$P(b) = 4/9 \times 3/8 = 12/72 = 1/6$$

FIGURE 2.16   Conditional event calculation.

2.11.5 Support Vector Machines

A support vector machine seeks to classify data with high orders of dimensionality as belonging to one set of information or another. As such, it has the potential of being applied to classifications of contexts as belonging to one supercontext or another. Such a case might be made by the conditional probabilistic reasoning mentioned earlier, where two flavors of context feature vectors are arriving over a given period of time; it is known that they are supercontexts of the feature vectors and are probably related (volcano versus tsunami), and a decision for an unknown new feature vector must be made about which superset an unknown feature vector might belong to. The support vector machine for superset classification and aggregation would do this classification by the generation of a maximal margin hyperplane, as seen in Figure 2.17.

In Figure 2.17, the approach is to the largest margin area that separates the data point values on the left and right sides of the graph. When creating the hyperplane, there are always two lines, one to the left and one to the right, that have the properties that (1) they do not misclassify data and (2) they seek to create the largest margin between the left and right lines, as shown in Figure 2.17.

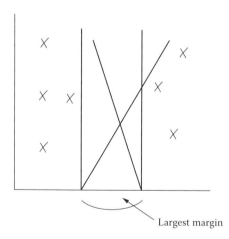

FIGURE 2.17   Example of various methods to find the largest margin among two sets of context feature vectors where the x's might be values of a single or multiple attributes in a context feature vector and an unclassified vector.

Intuitively, the property of having the largest margin reduces the number of classification errors that such an approach will generate and thus increases the model's generalization capabilities. More formally, the relationship of the classification error to the size of the margin is given by the following equation:

$$R \le R_e + \rho\left(\frac{h}{N}, \frac{\log(N)}{N}\right)$$

where

- R is the classification error;
- R_e is the training error;
- N is the number of training examples to generate the margin; and
- h is a descriptor of the complexity of the model, known as its *capacity*.

Capacity has an inverse relationship to margin. Models that have small margins have higher capacities because they can classify many more training sets of data. Of note is that as capacity increases, so does the rate of error generated in classification.

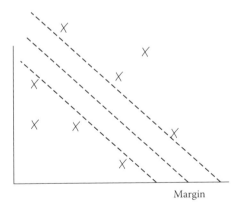

FIGURE 2.18   Another example of a relatively good classification margin where context features' attributes above the line belong to one supercontext (tsunami) and attributes below belong to another (volcano).

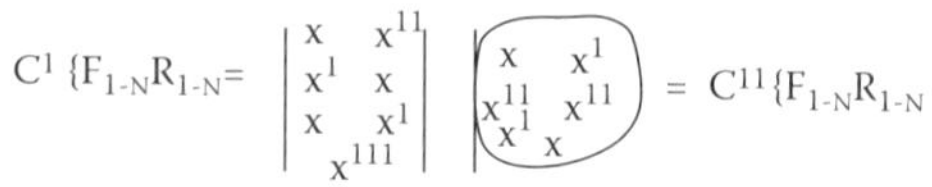

$$C^1\{F_{1\text{-}N}R_{1\text{-}N}= \begin{vmatrix} x & x^{11} \\ x^1 & x \\ x & x^1 \\ & x^{111} \end{vmatrix} \begin{pmatrix} x & x^1 \\ x^{11} & x^{11} \\ x^1 & \\ x^1 & x \end{pmatrix} = C^{11}\{F_{1\text{-}N}R_{1\text{-}N}$$

FIGURE 2.19 Classification of a feature vector $f\!\!\diamond$ as belonging to C' (volcano supercontext) or C'' (the supercontext for a tsunami) with associated contextual-processing rules R.

The application of this method in contextual reasoning might be to be able to create a classification model between two theoretical supercontexts from initial training or hypothetical data. As new contextual data feature vectors arrive, they can be classified as either (1) being new data that belong to a particular existing context; or (2) if a feature vector of data arrives, then the application of an support vector machine (SVM) could be utilized to classify a feature vector as a new context belonging to a set of contexts, as shown in Figure 2.19.

Application of support vector machines in conjunction with conditional probability could become the basis for (1) reasoning about theoretical models and classification of observed data into supercontexts describing thematic events based on the models, and (2) reasoning about the relation of at least two supercontexts, in this case the volcano supercontext and its relationship to the tsunami supercontext. The process of aggregation as discussed earlier can be the granular basis for mitigation of supercontexts where data are ambiguous or conflicting.

2.11.6 Clustering

Clustering is the process of classification of objects, in this case context feature vectors, into different *groups* that are closely related in some fashion. It is similar to that of support vector machines. The assumption of this method as found in previous methods is that generation of these groups will produce supercontexts of feature vectors that will have similar contextual rules and properties. Clustering could be done on multiple attributes found in a context vector, or there may be methods that could be developed for aggregation of clustering into a single model.

The process of developing clusters is essentially one of partitioning a context data attributes set into related subsets. The determination of membership is often done based on some concept of distance from the center of a cluster that an object belongs to. Clustering is a common technique for statistical data analysis.

The first step in clustering is to select a measure of distance that can be utilized to determine the similarity among contextual data attributes, and thus the supersets they might be aggregated into. The process of developing the clusters can start with selection of n random seed points, which then become the number of clusters found in the final model. Distance of objects is then calculated in some consistent fashion and an object is then assigned to a cluster based on which one it has a minimum distance from. After this initial step, a new center for a given contextual attributes cluster is found and the process iterates. Many methods of distance calculation might be considered for application of contextual data. Some potential functions might include the following:

- The Euclidean distance, which may be thought of as a straight-line distance between two contextual attributes values

- The Manhattan distance (also called *taxicab norm* or *1-norm*)

- The maximum norm

- The Mahalanobis distance based on different scales and correlations among variables

- Vector angular measures utilized for high-dimensional data

- The Hamming distance, which measures the minimum number of substitutions required to change one object to be the same as another object

Because contextual data feature vectors can be infinitely long, the simplest distance calculation might be to extend the concept of distance into a calculation of n dimension Euclidean distance given by the following equation.

Given a cluster's attribute f and a new unclassified contextual feature vector g,

$$f < a_1, a_2 \ldots a_n >$$

$$g < a_1, a_2 \ldots a_n >:$$

$$d = sqrt \; (Sum \; i \; to \; infinity \; (F_{ai} - G_{ai})),$$

where d is the distance measure between a known cluster and contextual cluster being considered for inclusion and thus aggregation.

This is a method that works as long as contextual feature vector attributes are numeric and ordered (discussed previously), but it is flawed in terms of its ability to analyze the context vectors proposed in the new model. Previously, context vectors have been defined to be composed of heterogeneous qualitative and quantitative types. For instance, consider two feature vectors given as

$$f < a_1, a_2, a_3, \ldots a_n >$$

$$g < a_1, a_2, a_3, \ldots a_n >$$

If the data types of these vectors are numeric for *all* attributes, then the mathematical calculation of distance is simple. However, earlier we defined feature vectors to be of numeric (quantitative) data and qualitative data. In order to calculate distance in this model, it might be proposed that the development of ontologies of concepts describing the hierarchical relationships among qualitative data be employed to handle nonordered and nonnumeric data. In such a scheme, distance could be measured based on the similarity or dissimilarity of concepts in the ontology. A number of methods could be developed to measure similarity. One of the simplest would be to calculate distance as a function of the number of concepts in the ontology that one has to traverse to get from concept a_1 to concept a_2. As an example, see Figure 2.20.

The calculation of dist (F, G), then, is best described by the following algorithm:

```
Start
Distance = 0
For I = 1 to n
     build ontological tree
```

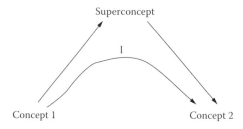

FIGURE 2.20 Calculation of distance between qualitative data in an ontology where *dist(concept1, concept2) = 1* via traversal through a superconcept.

```
End For
For i = 1 to n
        if attributetype (ai) = quantitative then
        Distance = Distance + Squareroot ((Fai - Gai)²)
        if attributetype (ai) = quantitative then
        Distance = Distance + DistanceOntology (Fai, Gai)
End For
End Start
DistanceOntology(Fai, Gai)
    Start at Fai
    Traverse to Ga,
    If not Gai then increment - count all nodes traversed
End DistanceOntology
```

2.11.7 Bayesian Techniques

Another technique that can be utilized to measure similarities in contexts and is thus useful for aggregation into supersets is Bayesian analysis. This method is based on conditional and marginal probabilities, and is similar to some of the principles of probability discussed earlier. This method reasons that if the probability of event x occurring is n, then the probability of y event occurring is m if we know they are somehow related. These probabilities relate the conditional and marginal probabilities of two random events. It is often used to compute what are called *posterior probabilities* given a given set of observations. For example, if contextual data suggest that a volcano in the ocean is occurring, then Bayesian methods can be used to compute the probability that a tsunami is going to occur given suggestive observations.

Application of Bayes' theorem is valid for the interpretations of probability. One philosophical notion behind Bayes is that it does have a conflicting approach to analysis compared to standard statistics. Specifically, one method that is called the *frequentist approach* assigns probabilities to random events according to the frequency of an event's occurrence. This approach is more similar to that of probabilistic reasoning, discussed earlier. The philosophy of Bayesian practitioners is to describe probabilities in terms of the beliefs and degrees of uncertainty. Thus, in a fashion it is somewhat similar to the methods of fuzzy set theory and incorporates more intuitively human understanding of what is being observed. In contextual classification this means that contexts can be classified as belonging to classes or sets of contexts based on the values of attributes in a context feature vector and the beliefs that this is probably a correct aggregation. This is a bit problematic, as humans will assign beliefs based on potentially conflicting interpretations of data and observation.

The equation for Bayes' theorem is given as

$$P(A|B) = \frac{P(B|A)P(A)}{P(B)}$$

The terms in Bayes' theorem describe the relationship of prior known probabilities and future calculated probabilities. A and B in this equation are events, observations, or values. Bayes' theorem in the general sense describes the way in which one's beliefs about observing A are derived from the observation of B.

The terms in the above equation are defined as the following:

- $P(A|B)$ is the conditional or posterior probability of A, given B. It is named as such because its value depends upon the value of B.

- $P(A)$ is referred to as the prior probability, sometimes also called the marginal probability of A. It is "prior" because its value does not consider B's value.

- $P(B)$ is the prior or marginal probability of B.

- $P(B|A)$ is the conditional probability of B given A.

As an example, if we observe that an attribute in feature vector $f\Diamond$ has a value of x for a given attribute, then we can make a probabilistic statement that this given vector is a member of context C_n. For example, let's define the following:

- $P(A|B)$ is the conditional probability that if an attribute value in a newly arrived context B is observed as n, then it belongs to context1, A.

- $P(A)$ represents the probability of a new context belonging to context1.

- $P(A')$ represents the probability of a new context not belonging to context2.

- $P(B)$ is the prior or probability of B, an existing context with a given attribute value.

- $P(B|A)$ is the conditional probability of B given A.

As an example of the Bayes' calculation, consider the following known observations. In a set of known supercontexts, S_1 and S_2, an attribute a_1 in a feature vector is known to contain the value 1 70% of the time and the value 2 30% of the time. When the a_1 attribute has the value 1 in a new context feature vector, then another attribute a_2 will contain the value 2 100% of the time. If a_1 contains the value 2, then a_2 will contain the value 1 or 2 50% of the time. If a new context feature vector arrives for similarity analysis and aggregation, and its value in attribute a_2 is found to be the value of 2, the question of the probability that this new context vector is similar to that of a set of contexts S_1 where $a_1 = 2$ and $a_2 = 1$ or 2 can be asked.

- $P(A|B)$ is the conditional probability that if an attribute value in a newly arrived context B is observed as n, then it belongs to the set A context.

- $P(A)$ represents the probability of a new context belonging to set A context based on attribute a_1 containing the value 2, which is .30.

- $P(A')$ represents the probability of a new context; the value 1 in attribute a_1 is .70.

- $P(B)$ is a randomly selected context containing the value 2 in a_2, which is given as $P(B|A)P(A) + P(B|A')P(A') = .5*.3 + 1*.70 = .85$.

- $P(B|A')$ is the probability that if $a_1 = 1$, then $a_2 = 1$, and is given as 1.

- $P(B|A)$ is the conditional probability of the new context containing in a_2 the value 1 or 2 is .5.

$$P(A|B) = .5*.3/.85 = .176$$

In this example, if we have only two sets of contexts, S_1, S_2, the low probability of the calculation .176 allows us to reason that the new context probably belongs to the S_2 set rather than the S_1 set by the logic that

$$P(S_2, \text{new context}) = 1 - P(S_2, \text{new context})$$

The Bayes' method can be applied for all sets of supercontexts to find which supercontext a newly arrived context vector has the highest probability of belonging to. Using this method to compare a newly arrived context against multiple sets of contexts makes it possible to put the context in the context set to which it has the highest probability of belonging; thus this approach can be utilized for aggregation.

2.11.8 Decision Trees

This method is a vehicle for classification commonly found in data mining. Starting with a small subset of information about a theme, the goal of a decision tree is to build a tree that will correctly classify all new types of contextual data based on the attributes in the vector. Typically this is done via Boolean reasoning and inequalities. Thus the method has a potential to handle certain types of data that do not lend themselves to previously discussed mathematical techniques.

In operations research, specifically in decision analysis, a decision tree is a decision support tool that uses a graph or model of decisions and their possible consequences, including chance event outcomes, resource costs, and utility. A decision tree is used to identify the strategy most likely to reach a goal. Another use of trees is as a descriptive means for calculating conditional probabilities.

In data mining and machine learning, a decision tree is a predictive model, that is, a mapping from observations about an item to conclusions about its target value. More descriptive names for such tree models are *classification tree* (discrete outcome) or *regression tree* (continuous outcome). In these tree structures, leaves represent classifications and branches represent conjunctions of features that lead to those classifications. The machine-learning technique for inducing a decision tree from data is called *decision tree learning*, or (colloquially) *decision trees*. Figure 2.21 shows a tree where values of N and M for a given attribute in a context vector can be utilized to arrive at a leaf node. In this case, leaf nodes could be constructed such that all feature vectors arriving at a given leaf node might then be candidates for aggregation into a supercontext. A problem with the technique is that the inclusion of n feature attributes in a context feature vector complicates the process of aggregation. However, the process might be extendible to classify very subjective things such as whether colors fall between other colors in subjective interpretation by humans.

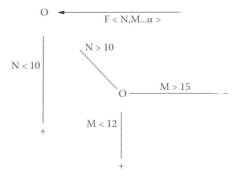

FIGURE 2.21   An example of a decision tree, classifying on the a_n and a_m attributes in context feature vector $f\diamond$.

2.12 OTHER TYPES OF REASONING IN CONTEXTS

Besides the above methods for reasoning about the similarity of contexts for aggregation into supersets, the ways that data are organized and stored logically after collection can suggest new methods that can be applied to reason about similarity in context feature vector data.

As an example, contextual queries are often expressed in spatial and temporal terms for the retrieval of information. In order to have such a query, data must first be organized into logical structures against which contextual processing may be applied. For this model, it may be useful to borrow a paradigm from geographic information systems and think of data as being logically (not physically) organized into thematic layers in a cube. Each layer may have the following logical condition: (1) a thickness, (2) no thickness, or (3) an undefined thickness. Thickness in this case can be utilized to assign a semantic meaning such as the degree of contextual importance in contextual processing of the data in a given thematic layer. Additionally, such layers could be thematic in nature where the ordering of layers from a given base layer could be an indicator of contextual similarity based on distance from a particular layer in the cube. Distance and similarity could be formulated into a series of equation models describing similarity and semantic relevance in processing. For example, in Figure 2.22, layer j may be the base layer for which we wish to describe similarity and semantic relevance. The distance equation

$$\text{distance}(i, j) = \text{similarity}(i, j)$$

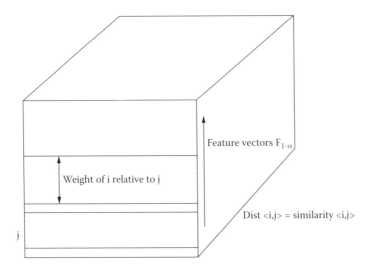

Feature vectors $F_{1-\alpha}$

Weight of i relative to j

Dist $<i,j>$ = similarity $<i,j>$

j

FIGURE 2.22   Model organization of context layer components where location in a cube might have semantic meaning (e.g., semantic importance, relation to, and so on) and distance has a degree of relation for the purpose of aggregation.

might couple into an equation for layer *i* that weights the distance with the thickness of layer *i* to become a measure of how relevant contextual-processing rules in layer *i* might be to the application of processing rules when processing layer *j* and/or degrees of potential aggregation that might occur among the layers. The appeal of such an approach is that it does have a spatial dimension that could be exploited to produce a more powerful contextual model paradigm; and because distance and thickness are only *logical concepts*, they could adaptively be modified depending on localized operational parameters as the model operates.

Besides the potential to aggregate data based on a logical storage model, many other types of models can potentially be developed for similarity analysis. Such models should hybridize previously discussed methods for conducting aggregation based on their similarity with novel logical abstractions describing and interrelating the four key dimensions of contextual processing: time, space, similarity, and impact.

2.13 CONTEXT QUALITY

Finally, any new or hybridized model of contextual processing needs to pay careful attention to the definition and measurement of quality metrics. Quality metrics will be key in the evaluation and selection of modeling to drive processing and probably need to be developed integrally as part of any model.

Definition and evaluation metrics should probably focus initially but not exclusively on key concepts that might be developed to describe quality. A few key criteria that could be considered are the development of the following concepts as key components of contextual-processing model quality:

- *Comprehension*: The degree that the information has been reduced in dimensionality to be comprehensible to human decision processes. Certain high-order problems will provide meaningless solutions in a context because the data may never converge to a consistent solution or a comprehensible solution.

- *Freshness*: The age of the data and therefore a possible indicator of how much confidence one may have in using the data for decision making or processing.

- *Usefulness*: Certain types of contextual feature vector data may not provide useful contextual processing because they are not relevant to a context theme. For example, if we are modeling the context of response to 9/11, then the color of the sky may have no relevance and therefore usefulness to this context.

- *Accuracy*: The degree to which the knowledge accurately models reality consistently.

- *Truth*: The degree to which it is known that the knowledge is accurate for a given instance. This concept should be interrelated to accuracy and might be a subcategory.

- *Ambiguity*: The degree to which we are *certain* that the knowledge represents the *real world* in an unambiguous fashion.

All of the previously discussed sections provide elements, concepts, and potential paradigms for contextual reasoning about similarity, organization of context data, semantic-processing grammars, and aggregation. Additionally, the notion that any model should equally emphasize the development of quality metrics at the same time a model is developed is emphasized. The next chapters will explore specific details and issues in logical reasoning about contexts, contextual data fusion, hyperdistribution to aid in the development of a global contextual model, powerful new paradigms for advance storage, query and similarity reasoning about the wide variety of information that could be part of a large contextual model, and finally a novel method that might suggest how security might be installed for contextual modeling based on the relevance of spatial and temporal relationships.

2.14 RESEARCH DIRECTIONS FOR GLOBAL CONTEXTUAL PROCESSING

This chapter introduced the paradigm of context and contextual processing. There is a tremendous amount of research that could be done and needs to be conducted to further the knowledge and definition of contextual processing and its model. Research can be done extensively in the area of reasoning about the similarity of context. This work needs to be done because data may be missing or incomplete in multiple contexts and thus need to be reconciled in order to drive processing. Most methods currently employed are rigorous in their application, and that probably should not be changed. However, the selection of data to which one will apply the method, or the interpretation of the result, could be based on contextually based methods where surrounding environmental data affect the similarity analysis process. For instance, in data mining, selection of the data set to which one will apply methods is critical, as is removal of outliers in the data set. Perhaps the contextual paradigm could be extended to contextually based data mining in the derivation of the test data set or the interpretation of the results. Success in this area could support more powerful similarity analysis methods. Perhaps one of the easiest areas for applying context would be fuzzy sets, which naturally lend themselves to reasoning about ambiguous situations.

Another area that could be researched is the semantics of processing and processing grammar development. A preliminary grammar was presented in this chapter purely as an example of how contextually based information might be married to processing to produce better responses to disasters. Finally, contextual processing works kind of like a black box that takes as input data about an event and contextual information. The output of that box is *contextual knowledge*. Contextual processing works on a similar model. It takes as its input more data, contextual knowledge (the output of the first black box), and produces as its output *contextual processing*. The working of both black boxes is undefined at the present.

The mechanisms inside the box, whether qualitative or quantitative, could be the subject of considerable further refinement and definition.

This chapter has introduced the concepts and initial modeling of contextual processing. However, contextual processing as mentioned previously will be implemented in an information system, which has the components of security, analysis, repository management, dissemination, and knowledge mining. Each of the following chapters looks at these key architectural components and how context and the concepts of contextual processing might be implemented. Most of the thinking presents new ideas and concepts that are not based on previous work. The next chapter will examine the application of logic and reasoning models in contextual processing. This topic is useful in the architecture of an IT system for the reason that conflicting information utilized in knowledge may actually be the same. Application of this topic can be utilized for better quality and confidence in the application of contextual data to processing.

REFERENCES

1. Rosemann, M., and J. Recker. (2006). "Context-aware process design: Exploring the extrinsic drivers for process flexibility." In T. Latour & M. Petit (eds.), *18th International Conference on Advanced Information Systems Engineering: Proceedings of Workshops and Doctoral Consortium*, pp. 149–158. Luxembourg: Namur University Press.
2. Schilit, B., N. Adams, and R. Want. (1994). "Context-aware computing applications." Paper presented at the *IEEE Workshop on Mobile Computing Systems and Applications* (*WMCSA'94*), Santa Cruz, CA, December.
3. Schilit, B. N., and M. M. Theimer. (1994). "Disseminating active map information to mobile hosts." *IEEE Network* 8 (5): 22–32.
4. Dey, A. K. (2001). "Understanding and using context." *Personal Ubiquitous Computing* 5 (1): 4–7.
5. Bolchini, C., C. A. Curino, E. Quintarelli, F. A. Schreiber, and L. Tanca. (2007). "A data-oriented survey of context models." *SIGMOD Rec. (ACM)* 36 (4): 19–26.
6. Schmidt, A., K. A. Aidoo, A. Takaluoma, U. Tuomela, K. Van Laerhoven, and W. Van de Velde. (1999). "Advanced interaction in context." In *International Symposium on Handheld and Ubiquitous Computing (HUC99)*, vol. 1707, pp. 89–101. Berlin: Springer LNCS.
7. Vert, G., S. S. Iyengar, and V. Phoha. (2009). "Security models for contextual based global processing: An architecture and overview." In *Cyber Security and Information Intelligence Research Workshop*. Oak Ridge, TN: Oak Ridge National Laboratory, ACM Digital Library.

8. Vert, G., S. S. Iyengar, and V. Phoha. (2009). "Defining a new type of global information architecture for contextual information processing." Paper presented at *IKE'09: The 2009 International Conference on Information and Knowledge Engineering*, Las Vegas, NV, July.

9. Vert G., and E. Triantaphyllou. (2009). "Security level determination using branes for contextual based global processing: An architecture." Paper presented at *SAM'09: The 2009 International Conference on Security and Management*, Las Vegas, NV, July.

10. Vert, G., J. Gourd, and S. S. Iyengar. (2009). "Integration of the visual authentication of spatial data with spatial-temporal class taxonomies for advanced spatial authentication modeling to create/pretty good security. Paper presented at the *2nd Cyberspace Research Workshop*, Center for Secure Cyberspace and USAF, Shreveport, LA, June.

CHAPTER 3

Calculus for Reasoning about Contextual Information

THEME

Previous chapters have introduced and discussed a model for contextual information processing, data types and considerations in such a model, analysis of contextual data vectors for aggregation, and the logic and underpinnings for why such a model is potentially useful. In this chapter, the considerations of the processing of contextual information using symbolic logic are presented. Such analysis and application of logic can lead to the development of powerful frameworks for reasoning about the truth of relationships between contextual data rules and their subsequent processing.

Logic has been studied since Aristotle's time; however, it was not until De Morgan's era that we learned to express any subject symbolically and define rules to manipulate those symbols. Over time, calculus for processing symbols has been further developed to address complex issues such as fuzzy inference, nonmonotonic reasoning, and specification of noneffects. In the following, we will dissect the representation of a context and its action rules and then discuss issues involving the deduction of inference from those action rules.

3.1 CONTEXT REPRESENTATION

Let us recall from Chapter 2 the representation of a context and its processing rules. A *context* is represented by a collection of attributes aggregated into a feature vector describing an event, and the *processing rules* of a context are the recommended actions for all expected scenarios for that context.

$$C = <a_1, a_2, \ldots, a_n>: R = \{r_1, r_2, \ldots, r_m\}$$

In the above, *C* is a context defined by a set of attributes $<a_1, a_2, \ldots, a_n>$ and *R* is the set of action rules for *C*. Rules in *R* represent consequence relationships for different premises defined by values of the attributes in *C*. An $r_i \in R$ is described as $r_i = C_i \Rightarrow w_i$, where C_i is a particular instance of *C* and w_i is the recommended action or set of actions for C_i. $\Rightarrow$ is an implication operator (e.g., $C_i \Rightarrow w_i$ is equivalent to saying, "If C_i, then w_i").

We make the following three observations on the above representation of context and its processing rules:

1. The attributes that define a context *C* are likely to be, in many cases, described using linguistic terms (e.g., "warm" weather) and hedges ("slightly" warm weather), and hence inferences using *R* need fuzzy processing.

2. The size of the vector $<a_1, a_2, \ldots, a_n>$ representing *C* may increase or decrease by future inclusion or removal of attributes of *C*, and that can sometimes affect the action rules in *R*, creating conflicts between rules. We need to apply some conflict resolution scheme in such situations.

3. The value of attributes of a context *C* may change over time, requiring an explicit description of the noneffects in the action rule.

For clarification, let us consider a hypothetical context that determines color codes used in airports to represent a terror threat level. Let us assume that associated context is defined by the following vector attributes:

$C = <$ (1) recent release of terrorist tapes, (2) reliable intelligence tips, (3) increased activity in the airport area, and (4) worldwide negative reaction to a U.S. policy $>$: $R = \{(C \approx C_1) \rightarrow$ Red Alert, $(C \approx C_2) \rightarrow$ Orange Alert, $(C \approx C_3) \rightarrow$ Green Alert$\}$

In the above, C_1, C_2, and C_3 represent three different combinations of the values of the features that define C. A fuzzy classifier can classify an instance of C into different classes. If it is later revealed that terrorists have a policy of launching attacks primarily on even dates, then we need to add attribute *date* in the context vector and redefine R as well. This is because a context that flagged a red alert earlier with respect to previous attributes may no longer need to do the same if the day has an odd date. Since inclusion of a new attribute may affect R, we need to use some appropriate calculus for it. Such a logical framework requires the use of nonmonotonic calculus to derive consequence relationships for contextual information. Also, unspecified temporal scenarios may lead to wrong inferences if all such possible future scenarios are not considered ahead.

3.2 MODUS PONENS

We assume that the readers are familiar with propositional, first-order, and fuzzy logic. The core inference rule for all of them is the *modus ponens*, as shown below:

$$P \Rightarrow Q$$
$$P$$
$$\therefore Q$$

However, they do not have similar expressiveness. Propositional logic's atomic unit is a proposition or a fact, and it has very limited expressiveness. Consider the following example:

If X flies on Monday, then X plots an attack. If X eats outside, then X flies. X eats outside and it is Monday. Does X plot an attack?

We can symbolize the above in propositional logic using the following:

Assuming $P = $ X flies, $Q = $ X plots an attack, $R = $ X eats outside, and $S = $ it is Monday, we can write

1. $(P \wedge S) \Rightarrow Q$
2. $R \Rightarrow P$
3. $R \wedge S$
4. R (from 3)
5. S (from 3)
6. P (from 2, *modus ponens*)
7. Q (from 1, 5, and 6, *modus ponens*), that is, X plots an attack.

However, if we change the above in the following way, propositional logic can no longer make this inference:

> If terrorists fly on Monday, then they plot an attack. X is a terrorist. If X eats outside, X flies. X eats outside and it is Monday. Does X plot an attack?

First-order logic (FOL) with the use of propositions and quantifiers can solve it. Unlike propositional logic, which validates arguments only according to their structure, FOL considers the deeper structure of propositions. The following shows the inference using forward chaining in FOL for the above problem:

1. $(\forall x)(\forall y)$ Terrorist(x) $\wedge$ Flies(x) $\wedge$ Monday(y) $\Rightarrow$ PlotanAttack(x)

2. Terrorist(X)

3. Eatsoutside(X) $\Rightarrow$ Flies(X)

4. Eatsoutside(X) $\wedge$ Monday (Y)

5. Terrorist(X) $\wedge$ Flies(X) $\wedge$ Monday(Y) $\Rightarrow$ PlotanAttack(X) (from 1, Universal Elimination)

6. Eatsoutside(X) (from 4)

7. Monday (Y) (from 4)

8. Flies(X) (from 3)

9. PlotanAttack(X) (from 2, 5, 7, and 8, *modus ponens*), that is, X plots an attack.

Fuzzy logic generalizes *modus ponens* in the following way:

$$P \Rightarrow Q$$

$$P' \therefore Q'$$

Here, P' and Q' may not be exactly P and Q, respectively. We can use the compositional rule of inference to derive their implications.

3.3 FUZZY SET AND OPERATIONS

A fuzzy set is any set that does not require a strictly crisp membership, but instead allows its members to have various degrees of membership ranging from 0 to 1.

3.3.1 Union

The union operation is the OR equivalent of Boolean algebra. Union of two fuzzy sets with membership functions μ_A and μ_B is a fuzzy set with the following membership function:

$$\mu_{A \cup B} = \max (\mu_A, \mu_B)$$

3.3.2 Intersection

The intersection operation is the AND equivalent of Boolean algebra. Intersection of two fuzzy sets with membership functions μ_A and μ_B is a fuzzy set with the following membership function:

$$\mu_{A \cap B} = \min (\mu_A, \mu_B)$$

3.3.3 Complement

The complement operation is the NOT equivalent of Boolean algebra. The complement of a fuzzy set is a set with the following membership function:

$$\neg \mu_A = 1 - \mu_A$$

3.3.3.1 De Morgan's Law

$$\neg (A \cap B) = \neg A \cup \neg B$$

$$\neg (A \cup B) = \neg A \cap \neg B$$

3.3.3.2 Associativity

$$(A \cap B) \cap C = A \cap (B \cap C)$$

$$(A \cup B) \cup C = A \cup (B \cup C)$$

3.3.3.3 Commutativity

$$(A \cap B) = (B \cap A)$$

$$(A \cup B) = (B \cup A)$$

3.3.3.4 Distributivity

$$A \cap (B \cup C) = (A \cap B) \cap (A \cap C)$$

$$A \cup (B \cap C) = (A \cup B) \cap (A \cup C)$$

We will later explain how the fuzzy inference scheme fits into our proposed framework. The details of fuzzy set theory are beyond the scope of this chapter. A recommended reference for interested readers is *Fuzzy Set Theory and Its Applications* by Hans-Jürgen Zimmermann [1].

3.4 CONTEXTUAL INFORMATION AND NONMONOTONIC LOGIC

First-order logic satisfies the following property:

$$(\Pi \Rightarrow \varphi) \Rightarrow ((\Pi \cup \delta) \Rightarrow \varphi'), \text{ such that } \varphi' = \varphi \text{ for any } \delta,$$

where Π is the set of given premises, φ is a conclusion derived from Π, and δ is a premise not present in Π.

Basically, it states that if a conclusion is made based on a given premise set, the addition of new premises in that premise set will not reduce that conclusion. The above property is formally known as *monotony*.

Nonmonotonic logic [2] is not restricted by the property of monotony. It represents a family of formal frameworks for models involving tentative decisions with respect to known facts and often requires specific types of quantification of predicates that cannot be done using first-order logic.

3.4.1 Conflicts in Conclusions

If a new premise is added to the premise set of a nonmonotonic logical framework, then the following two types of conflicts may occur:

1. Conflicts between previous tentative conclusions and the newly added premise. This is described by

$$\Pi \Rightarrow \varphi; \text{ however, } (\Pi \cup \delta) \Rightarrow \varphi', \text{ such that } \varphi' = \neg \varphi$$

where Π is the set of given premises, φ is a conclusion derived from Π and δ is a premise not present in Π, and φ' is a conclusion derived from $\Pi \cup \delta$. Such conflicts can be eliminated by removing $\Pi \Rightarrow \varphi$ from the conclusions.

2. Conflicts between two contradicting conclusions deduced from the same framework. This is described by

If $(x \Rightarrow \varphi) \in \Pi$ and $(x \Rightarrow \varphi') \in \delta$, such that $\varphi' = \neg \varphi$, then $(\Pi \cup \delta) \Rightarrow$

$(x \Rightarrow \varphi$ and $x \Rightarrow \varphi')$,

where $x \Rightarrow \varphi$ is a premise in a set of premise Π, and $x \Rightarrow \varphi'$ is a premise in a set of premise δ.

Inclusion of "Increased activity in the airport area⇒Enhance security activity" in a set of premises containing "Increased activity in the airport area⇒Terrorist activity" will introduce a conflict of this latter kind, as depicted in Figure 3.1.

This type of conflict can be solved using two different approaches resulting in possibly two different conclusions, each having its own justification. The first approach discards both the conflicting conclusions, and the second one picks up one or the other. Picking up one of the solutions using additional knowledge can be appropriate in some contexts, where ignoring both the conclusions may have different real-world implications. For instance, in Figure 3.1, raising the red alert might be the right conclusion since the cost of "raising the red alert" will be the least had a wrong conclusion been made.

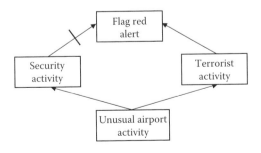

FIGURE 3.1   Conflict in conclusions.

To resolve conflicts of type (2) in the perspective of contextual premises, we recommend the use of the *prioritizing model,* where preference of one formula over the other is predefined to resolve any conflict. The preferences in a preferential model will be domain dependent, making sure that the cost of the resolved conclusion is the minimum over all the options. We discuss these in detail in the following section.

3.4.2 Default Theory

In first-order logic, all the premises describe that something is either true or false. However, in the contextual perspective, we may often have to apply commonsense reasoning such as stating that something is true by default. A classic example is the reasoning "Birds typically fly" instead of "Birds fly." We need some new type of formalism to express and process such default reasoning. We will base our discussion on the formalism that Raymond Reiter [3] introduced.

3.4.2.1 Default

A default consists of a set of prerequisites P, a justification R, and a conclusion Q where P implies Q if R is consistent with current belief and is expressed as $\frac{P:R}{Q}$.

3.4.2.2 Default Theory

A default theory is a pair $\langle D, W \rangle$ where D is a finite set of defaults and W is a finite set of formulae representing the initial set of beliefs. The *initial set of beliefs* is continually updated with the new conclusions we make and used as the *current belief* to determine the consistency of justifications in D. If we express $\langle D, W \rangle$ as $\langle \frac{P:R}{Q}, W \rangle$, then we can describe this as follows: if P is true and R is consistent with W, then Q. The default rule "Birds typically fly" can be formalized by the following default theory:

$$\left\langle \left\{ \frac{\text{Bird}(X): \text{Flies}(X)}{\text{Flies}(X)} \right\}, \{\neg\text{Flies}(\text{Ostrich})\} \right\rangle$$

This rule means that if X is a bird and if the assumption that X flies does not conflict with the belief {¬Flies(Ostrich)}, then we can conclude that X flies. According to this, Bird(Parrot) implies Flies(Parrot) because Flies(Parrot) does not conflict with the background theory. However, Bird(Ostrich) does not imply Flies(Ostrich) because it conflicts with the fact ¬Flies(Ostrich) in the background theory.

The consequence Q of a default rule adds new belief to the existing belief W, as shown below.

$$\left\langle \left\{ \frac{P{:}R}{Q},... \right\}, W \right\rangle \qquad \left\langle \left\{ \frac{P{:}R}{Q},... \right\}, W \cup \{Q\} \right\rangle$$

Thus subsequent defaults can be applied to the current belief updated by already applied defaults. When a default theory is such that no other default can be applied to it, the theory is called an *extension of the default theory*. Obviously the following question arises: can such an extension be affected by the order of the application of the defaults? The answer is yes. This is explained below by the classical Nixon diamond example that has two extensions.

$$\left\langle D = \left\{ \frac{\text{Republican }(X){:} \neg\text{Pacifist}(X)}{\neg\text{Pacifist}(X)}, \frac{\text{Quaker}(X){:} \text{Pacifist}(X)}{\text{Pacifist }(X)} \right\} \right.$$

$$\left. W = \{\text{Republican(Nixon), Quaker(Nixon)}\} \right\rangle$$

In the above, $|D|=2$ (i.e., D has two elements), and for both the elements their corresponding Rs (i.e., Pacifist(Nixon) and ¬Pacifist(Nixon)) are consistent with W={Republican(Nixon), Quaker(Nixon)}, although they are not consistent with each other. Thus, application of the first default creates the following extension:

$$\left\langle D = \left\{ \frac{\text{Republican }(X){:} \neg\text{Pacifist}(X)}{\neg\text{Pacifist}(X)}, \frac{\text{Quaker}(X){:} \text{Pacifist}(X)}{\text{Pacifist }(X)} \right\}, \right.$$

$$\left. W = \{\text{Republican(Nixon), Quaker(Nixon) } \neg\text{Pacifist (Nixon)}\} \right\rangle$$

The second default cannot be applied to this theory.

However, if the default is applied first, we arrive at the following extension:

$$\left\langle D = \left\{ \frac{\text{Republican (X): } \neg\text{Pacifist(X)}}{\neg\text{Pacifist(X)}}, \frac{\text{Quaker(X): Pacifist(X)}}{\text{Pacifist (X)}} \right\} \right.$$

$$W = \{\text{Republican(Nixon), Quaker(Nixon) } \neg\text{Pacifist(Nixon)}\} \left. \right\rangle,$$

where the first default cannot be applied.

Since the order of the application of defaults can have totally different extensions, how does one entail a given formula? There are two primary approaches: skeptical entailment and credulous entailment. Under skeptical entailment, a formula is entailed by a default theory if it is entailed by all its extensions created by applying the defaults in different orders. On the other hand, credulous entailment entails a formula if it is entailed by one or more of the extensions of the default theory.

3.4.3 Entailment in a Contextual Case

In the contextual model case, where we may deal with problems like disaster management or terror activity monitoring, we may want to be too precautious and adopt credulous entailment. However, not all the extensions of a default theory may have the same significance, and if we have a prioritized order of the application of the defaults, we may avoid unnecessary entailments of formulae and thus make the solution more efficient. *Prioritized default theory* [4] formalizes this solution.

3.4.3.1 Prioritized Default Theory

A prioritized default theory is a triple $\langle D, W < \rangle$, where D is a finite set of defaults, W is a finite set of formulae representing the initial set of beliefs, and $<$ is a partial ordering on D expressed as $\delta_n < \ldots \delta_2 < \delta_1$, such that δ_i is applied to the default theory before δ_{i+1} to derive the extension that determines the entailments of the formula. The priorities are, of course, determined by the context, and one requires appropriate domain knowledge to assign them.

3.5 SITUATION CALCULUS

Situation calculus was introduced by John McCarthy as a logic formalism for dynamic domains initially in 1963 and was improved thereupon in 1980 in an attempt to handle posteffect scenarios known as the *frame*

problem [5,6]. However, his formalism fails to formalize special case scenarios. Below we define the frame problem, state McCarthy's approach of circumscription toward its solution and show why circumscription fails under some temporal scenarios such as the Yale shooting problem, and finally describe Ray Reiter's solution to the situation calculus introduced in 1991 [7] that eliminates the limitations encountered by circumscription.

3.5.1 Frame Problem

In contextual representation, we describe the set of actions required for a specific context and we only consider the effects of those actions. In a logical framework, if we do not specify the noneffects of those actions, we may not be able to process them. For example,

1. $\forall x \forall a\ P(x,a) \Rightarrow Q(x,a)$

2. $\forall x \forall b\ R(x,b) \Rightarrow S(x,b)$

3. $P(X,A)$

4. $R(X,B)$

For the above scenario, $Q(X,A)$ may or may not be true depending on whether $R(X,B)$ has any effect on it. We require explicitly specifying the noneffects of an action to overcome this problem. For instance, explicit inclusion of the following ensures that $Q(X,A)$ is true.

1a. $\forall x \forall a \forall b\ Q(x,a) \Rightarrow (R(x,b) \Rightarrow Q(x,a))$

2a. $\forall x \forall a \forall b\ S(x,b) \Rightarrow (P(x,a) \Rightarrow S(x,b))$

In general, there can be a large number of noneffects for any action, making such representations impractical and error-prone. This problem is known as the *frame problem*.

3.5.2 Circumscription and the Yale Shooting Problem

John McCarthy applied circumscription to address the frame problem [3, 4]. Circumscription assumes that every action is associated with minimal change (i.e., things are as expected unless otherwise specified). However, minimal change may often misrepresent temporal changes, as we will see in the Yale shooting problem scenario described below.

Steve Hanks and Drew McDermott introduced this example scenario of nonmonotonic temporal reasoning. The scenario involves an imaginary character, Fred, who may or may not be killed by gunfire. There is a sequence of situations. In the initial situation, a gun is not loaded and Fred is alive. Loading the gun, waiting, and then shooting Fred are supposed to kill Fred. If we use circumscription and assume a minimal temporal change, then the fact that the gun is loaded and "Fred dies" and the gun may become somehow unloaded and "Fred survives" both satisfy. Two such minimal change scenarios are shown below:

Given:

$$alive(0)$$

$$\neg loaded(0)$$

$$true \Rightarrow loaded(1)$$

$$loaded(2) \Rightarrow \neg alive(3)$$

In the above, at time 0, Fred is alive and gun is not loaded. At time 1, the gun is loaded. At time 2, if the gun is loaded (and the gun is shot), then at time 3 Fred is dead.

Scenario 1:			
alive(0)	alive(1)	alive(2)	$\neg$alive(3)
$\neg$loaded(0)	loaded(1)	loaded(2)	loaded(3)
Scenario 2:			
alive(0)	alive(1)	alive(2)	alive(3)
$\neg$loaded(0)	loaded(1)	$\neg$loaded(2)	$\neg$loaded(3)

In scenario 1, state of loaded() and alive each changed once (for a total of two changes); and in scenario 2, state of loaded() changed twice, but isn't the unloading at time 2 an extra assumption—it should not change automatically? Thus, the standard circumscription for the frame problem fails to formalize dynamic scenarios such as those described in the Yale shooting problem and requires the situation calculus formalism.

In the above scenario there are two actions, *load* and *shoot*, and two "fluents," *alive* and *loaded*. A *fluent* is a state that can change its truth value over time if some action takes place.

Loading the gun only changes the value of *loaded* and not the value of *alive*, introducing the framing problem that requires derivation of the *noneffects* of actions. However, as mentioned earlier, describing all such noneffects can be impractical due to their large number. In situation calculus, the frame problem is addressed by successor state axioms that describe all possible ways that a fluent can change its state by considering different time points and relating every fluent to time.

3.5.3 Formalism

We describe Pirri and Reiter's [8] formalism of situation calculus in this section, which takes a more comprehensive approach to the frame problem.

Situation calculus is defined by three elements: actions, situations, and fluents.

3.5.3.1 Action

An action is a quantifiable event that has a consequence. An action may or may not be executed depending on the preconditions. For example, it is impossible to shoot a gun before it is loaded. This is described through

$$Poss(a,s)$$

where a is an action, s is a situation, and $Poss(a,s)$ denotes the executability of a at s. For example, $Poss(Shoot(o),s) \Leftrightarrow loaded(o,s)$ states that o can shoot at a situation s if and only if o is loaded at s.

3.5.3.2 Situation

A situation represents a finite sequence of actions. A new situation results from a previous situation and an action. The initial situation is typically denoted by S_0, and the subsequent situations are derived using the following: action×situation⇒situation.

3.5.3.3 Fluent

A fluent, as described before, is a state of the world that can change its value. Whenever an action may take place in a situation, it is required that its effect on the fluent be specified. For instance, $Poss(Shoot(o),s) \Rightarrow alive(Fred, do(Shoot(o),s))$.

3.5.4 The Successor State Axioms

We have mentioned earlier that solving the frame problem using standard circumscription may lead to wrong results in situation calculus. To solve this problem, successor state axioms have been used that enumerate all possible states that a specific fluent can assume. Using the notations of Pirri and Reiter, we can generalize such effect on a fluent F by formulating both positive (F^+) and negative (F^-) effect axioms as

$$Poss(a, s) \land F^+ \Rightarrow F \text{ and } Poss(a, s) \land F^- \Rightarrow \neg F$$

This formulation does not leave any space for any noneffect scenario that could appear otherwise as explained above.

3.6 RECOMMENDED FRAMEWORK

The core of the framework is a fuzzy inference model, as shown in Figure 3.2.

3.6.1 Fuzzy Inference Scheme

The core of the framework is a fuzzy inference scheme. Recall that in $C = <a_1, a_2, ..., a_n>$: $R = \{r_1, r_2, ..., r_m\}$, $<a_1, a_2, ..., a_n>$ is an attribute vector that defines context C (which determines premises) and R is the set of action rules $= \{r_1, r_2, ..., r_m\}$.

Most likely, the set R and C will be described in fuzzy language and will require fuzzy inference processing. As a first step to fuzzy inference processing, we need to assign fuzzy values to each attribute in C and also

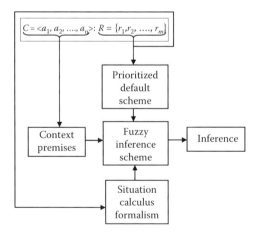

FIGURE 3.2 Basic framework for contextual information processing.

in the action rules in R. A convenient representation could be a tuple (i,j), denoting that the value of i is j. For instance, we can define C_1, C_2, and C_3 in the following way:

$C = \,<$(1) recent threat by terrorists, (2) poor intelligence tips, (3) increased activity in the airport area, and (4) reaction against a U.S. policy$>$: $R = \{(C \approx C_1) \rightarrow$ Red Alert, $(C \approx C_2) \rightarrow$ Orange Alert, $(C \approx C_3) \rightarrow$ Green Alert$\}$

where

$\quad C_1 = \{(1, .5), (2, .7), (3, .5), (4, .5)\}$

$\quad C_2 = \{(1, .3), (2, .5), (3, .5), (4, 0)\}$

$\quad C_3 = \{(1, .3), (2, .5), (3, .5), (4, 0)\}$

Appropriate membership values are assigned to different attributes empirically or axiomatically using analysis, experience, and intuition. Usually a lookup table is built or a function is designed to defuzzify a hedge or qualifier into a fuzzy membership value. Once the input is ready for fuzzy inference, we can apply fuzzy operators on them and conclude.

3.7 EXAMPLE

Let us revisit the tsunami example, considering three attributes for the context C as the following:

$\quad C = \,<$seismographic data, satellite imagery, natives' intuition$>$: R

The following steps are required to support this scheme using the proposed framework:

"Highly increased wave temperature moderately implies a tsunami wave."

"Very calm wave strongly implies not a possible tsunami wave."

While we formulate and prioritize defaults and resolve the frame problem, we will temporarily ignore the hedges (such as *highly*, *moderately*, *very*, *strongly*, etc.). We will consider them during the fuzzy inference process.

3.7.1 Prioritize Defaults

If the definition of C is extended over time, we may have conflicting conclusions, as discussed before. To resolve this, we will apply the prioritizing default theory. For instance, we may have two different extensions depending on the order of defaults in the following scenario. Given:

$$\left\langle D = \left\{ \frac{\text{Calm(wave)}: \neg\text{Tsunami(wave)}}{\neg\text{Tsunami(wave)}}, \frac{\text{Warm(wave)}:\text{Tsunami(wave)}}{\text{Tsunami(wave)}} \right\}, \right.$$

$$\left. W = \{\text{Tsunami(Indian wave)}\} \right\rangle$$

If we find the following

1. It is a very calm Pacific wave, but it has recently become very warm. Then W may include Tsunami(Pacific wave) or ¬Tsunami(Pacific wave), depending on which of the defaults is evaluated first.

We set the following priority of the defaults, considering a false positive warning better than a false negative warning:

$$\frac{\text{Warm(wave)}:\text{Tsunami(wave)}}{\text{Tsunami(wave)}} > \frac{\text{Warm(wave)}: \neg\text{Tsunami(wave)}}{\neg\text{Tsunami(wave)}}$$

Then the extension becomes

$$\left\langle D = \left\{ \frac{\text{Warm(wave)}: \text{Tsunami(wave)}}{\text{Tsunami(wave)}}, \frac{\text{Calm(wave)}: \neg\text{Tsunami(wave)}}{\neg\text{Tsunami(wave)}} \right\}, \right.$$

$$\left. \{\text{Tsunami(Indian wave)}, \text{Tsunami(Pacific wave)}\} \right\rangle$$

3.7.2 Resolve the Frame Problem

It may so happen that there had been signs of tsunami waves; however, further investigation of the wave reveals no such significant risk. If we say to apply the circumscription method and assume a minimum change, the risk will not be eliminated. This may lead to the use of wrong action rules. We resolve this problem in the following way:

Given:

1. A warm wave implies not a possible tsunami wave if further investigation of the wave reveals no such significant risk.

2. A warm wave implies definitely a possible tsunami wave if later a high seismographic warning is raised in the region.

We can apply successor state axioms and add the following:

Poss(Tsunami(Pacific wave), Tsunami risk(pacific wave)) ∧ {Seismographic warning (Pacific)} ⇒ ¬Tsunami(Pacific, do(earthquake(Pacific), Tsunami risk(Pacific wave))

Poss(Tsunami(Pacific wave), Tsunami risk(Pacific wave)) ∧ {No tsunami risk (Pacific)} ⇒ ¬Tsunami(Pacific, do(warm(Pacific), Tsunami risk(Pacific wave))

Thus the frame problem will be taken care of for the scenarios encountered afterward.

3.7.3 Fuzzy Inference

Given a circumstance described as "It is a very calm Pacific wave, but it has recently become very warm. A slight seismographic warning is raised," if we apply prioritized defaults and resolve the frame problem, we conclude that a tsunami may happen upon an earthquake. However, is that an absolute possibility? If not, what is the likelihood of it happening? We will apply fuzzy inference to compute this likelihood.

Fuzzification is the important first step in the fuzzy inference process that determines the degree of membership of an input to the associated set. We can either do a table lookup or apply a membership evaluation function for it. In the example of Figure 3.3, we apply arbitrary evaluation functions for explaining the process.

Given:

1. A high increase in wave temperature and a significant change in the electromagnetic (EM) spectrum moderately imply a tsunami wave.

2. A very calm wave strongly implies not a possible tsunami wave.

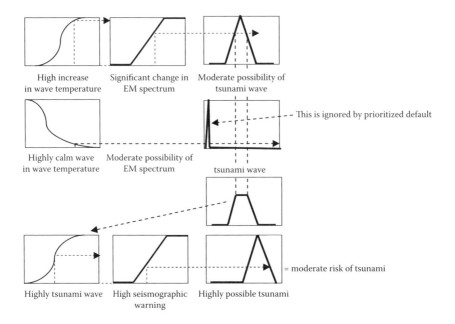

FIGURE 3.3   Fuzzy inference.

3. A high tsunami wave implies not a possible tsunami if further investigation of the wave reveals no such significant risk.

4. A high tsunami wave implies a highly possible tsunami if later a high seismographic warning is raised in the region.

We process the inference for the following context:

It is a very calm Pacific wave, but it has recently become very warm and there has been some change in the EM spectrum. A light seismographic warning is raised.

In Figure 3.3, first we compute the degree to which it is a tsunami wave and retain its prioritized value. Later we use that prioritized value to evaluate the degree of tsunami risk for the post effects. Our inference is that for the given context, there exists a moderate risk of tsunami.

3.8 CONCLUSION

Different issues involving automated contextual information processing can be resolved using several competitive approaches. We propose a specific solution using a combination of fuzzy inference, default logic, and situation calculus. Default logic provides the nonmonotonic property support that we may need as we update the contextual information-processing system, such as its action rules. We also suggest prioritizing the defaults to resolve any potential conflicts that may arise while computing the extension in a default theory. To formalize temporal changes we recommend the use of situation calculus, and to solve the frame problem we recommend the use of the successor state axioms.

REFERENCES

1. Zimmermann, H.-J. (1985). *Fuzzy set theory and its applications*, 4th ed. New York: Springer.
2. Hanks, S., and D. McDermott. (1987). "Nonmonotonic logic and temporal projection." *Artificial Intelligence* 33 (3): 379–412.
3. Reiter, R. (1980). "A logic for default reasoning." *Artificial Intelligence* 13: 81–132.
4. M. Gelfond and T.C. Son. (1998). Prioritized default theory. In *Selected Papers from the Workshop on Logic Programming and Knowledge Representation*. Springer. 164–223.
5. McCarthy, J., and P. Hayes. (1969). "Some philosophical problems from the standpoint of artificial intelligence." *Machine Intelligence* 4: 463–502.
6. McCarthy, J. (1980). "Circumscription: A form of non-monotonic reasoning." *Artificial Intelligence* 13: 27–39.
7. Reiter, R. (1991). The frame problem in the situation calculus: A simple solution (sometimes) and a completeness result for goal regression. In: V. Lifschitz, Editor, *Artificial Intelligence and Mathematical Theory of Computation: Papers in Honor of John McCarthy*, Academic Press, San Diego, CA. 359–380.
8. Pirri, F., and R. Reiter. (1999). "Some contributions to the metatheory of the situation calculus." *Journal of the ACM* 46 (3): 325–361.

CHAPTER 4

Information Mining for Contextual Data Sensing and Fusion

THEME

Previous chapters have examined logic and the underpinning of the new model for contextual processing. For such a model to operate on a global scale, it must deal with issues of hyperdistribution of information across n number of platforms and information repositories. Aspects of how hyperdistribution of contextual information and advanced repository models are presented in subsequent chapters. Once data are collected and aggregated by similarity into supercontexts, they can be sent to multiple consumers. Because consumers may in theory have multiple sources of contextual information and/or find information useful from other consumer repositories, there is a need to examine how distributed data mining (DDM) for contextual processing might occur. Such mining can lead to new contextual information utilized to affect contextual processing and further aggregation and analysis of supercontexts into hyperknowledge.

This chapter provides an overview of data mining (DM) in its early sections. It then discusses issues with DDM and discusses an architecture, the Knowledge Grid (K-Grid) for accomplishing distributed mining, proposed by Talia et al. Finally, a discussion is presented about how such methods might be merged with context-sensing, intelligent, knowledge-based derivation.

This chapter presents keystone knowledge that is built upon in subsequent chapters, where the hyperdistribution of contextual-processing information is examined and an advanced model for repository management based on the relations of information is developed and presented.

4.1 DATA-MINING OVERVIEW

The term *data mining* has been stretched beyond its limits to mean any form of data analysis. Data mining—or *knowledge discovery in databases* (KDD), as it is also known—is the extraction of nontrivial implicit, previously unknown, and potentially useful information from a database. This encompasses a number of different technical approaches, such as clustering, data summarization, learning classification rules, finding dependency in networks, analyzing changes, and detecting anomalies.

Data mining is the search for relationships and global patterns that exist in large databases but are "hidden" behind a vast amount of data. An example may be the relationship between data of patients and their medical diagnosis. These relationships represent valuable knowledge about the database and the objects in the database and, if the database is a faithful mirror, of the real world registered by the database.

Data mining refers to "using a variety of techniques to identify nuggets of information or decision-making knowledge in bodies of data, and extracting these in such a way that they can be put to use in the areas such as decision support, prediction, forecasting and estimation. The data is often voluminous, but as it stands, of low value, as no direct use can be made of it; it is the hidden information in the data that is useful."

Basically, data mining is concerned with the analysis of data and the use of software techniques for finding patterns and regularities in sets of data. It is the computer which is responsible for finding the patterns by identifying the underlying rules and features in the data. The idea is that it is possible to strike gold in unexpected places as the data-mining software extracts patterns not previously discernable or so obvious that no one has noticed them before.

Data-mining analysis tends to work from the data up, and the best techniques are those developed with an orientation toward large volumes of data, making use of as much of the collected data as possible to arrive at reliable conclusions and decisions. The analysis process starts with a set of data and uses a methodology to develop an optimal representation of the structure of the data, during which time knowledge is acquired. Once knowledge has been acquired, this can be extended to larger sets of data working on the assumption that the larger data set has a structure similar to the sample data. Again, this is analogous to a mining operation where large amounts of low-grade materials are sifted through in order to find something of value.

The phases depicted start with the raw data and finish with the extracted knowledge which was acquired as a result of the following stages:

- *Selection*: Selecting or segmenting the data according to some criteria (e.g., all those people who own a car). In this way, subsets of the data can be determined.

- *Preprocessing*: This is the data-cleansing stage where certain information is removed which is deemed unnecessary and which may slow down queries. For example, it is unnecessary to note the sex of a patient when studying pregnancy. Also the data are reconfigured to ensure a consistent format as there is a possibility of inconsistent formats because the data are drawn from several sources (e.g., sex may be recorded as form and also as 1 or 0).

- *Transformation*: The data are not merely transferred across but also transformed, in that overlays may be added such as the demographic overlays commonly used in market research. Thus the data are made usable and navigable.

- *Data mining*: This stage is concerned with the extraction of patterns from the data. A pattern can be defined as given a set of facts (data) F, a language L, and some measure of certainty C, a pattern is a statement S in L that describes relationships among a subset F_s of F with a certainty C such that S is simpler in some sense than the enumeration of all the facts in F_s.

- *Interpretation and evaluation*: The patterns identified by the system are interpreted into knowledge, which can then be used to support human decision making, for example prediction and classification tasks, summarizing the contents of a database, or explaining observed phenomena.

The performance analysis of DM systems has remained as a challenge over the last several years. Reasons for this concern are (1) the size and distributed nature of input data, (2) the spatial and temporal complexity of DM algorithms, and (3) the quasi-real-time constraints imposed by many applications. When it is not possible to control the performance of DM systems, the actual applicability of DM techniques is compromised.

Data mining is the science of extracting useful and nontrivial information from the huge amounts of data collected from many and diverse fields of science, business, and engineering. Due to its relatively recent development, DM still poses many challenges to the research community. New methodologies are needed in order to mine more interesting and specific information from the data, new frameworks are needed to harmonize more effectively all the steps of the mining process, and new solutions will have to manage the complex and heterogeneous sources of information that are today available for the analysts.

4.2 DISTRIBUTED DATA MINING

Data-mining technology is not only composed by efficient and effective algorithms, executed on stand-alone kernels. Rather, it is constituted by complex applications articulated in the nontrivial interaction among hardware and software components, running on large-scale distributed environments. For a growing number of applications, DDM is an important tool.

4.2.1 Motivation for Distributed Data Mining

Some examples of data collected in a distributed fashion are sales points of a large chain such as Walmart, or the branches of a bank or the census offices in a country. These data are typically too big to be gathered at a single site. For privacy issues, they are moved within a limited set of alternative sites. In this situation, the execution of DM tasks typically

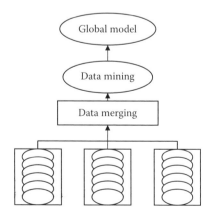

FIGURE 4.1   The data-driven approach for DDM systems.

involves the decision of distributing data to different locations. Typical examples are earth-observing systems (EOS) [1], such as satellites sending their observational data to different earth stations, or high-energy physics experiments that produce huge volumes of data. In these cases, data can be replicated in more than one site and repositories can have a multitier hierarchical organization [2]. Problems of replica selection and caching management are typical in such applications.

4.2.2 DDM Systems

Some definitions of DDM systems are discussed here by analyzing three different approaches. They pose different problems and have different advantages. Existing DDM systems can, in fact, be classified in one of these approaches.

4.2.2.1 A Data-Driven Approach

The simplest model for a DDM system takes into account only the distributed nature of data. Since in this system the focus is solely on the location of data, this model is referred as *data driven*.

In this model, data are located in different sites which may or may not have any computational capability. The only requirement is the ability to move the data to a central location in order to merge them and then apply sequential DM algorithms. The output of the DM analysis (i.e., the final knowledge models) is then either delivered to the analyst's location or accessed locally where it has been developed.

Data can be partitioned horizontally (i.e., different records are placed in different sites) or vertically (i.e., different attributes of the same records are distributed across different sites). Also, the scheme itself can be distributed (i.e., different tables can be placed in different sites). It is necessary to adopt a proper merging strategy when gathering data.

4.2.2.2 A Model-Driven Approach

The second approach is the one called *model driven*, and it is depicted in Figure 4.2. Here, each portion of data is processed locally (in its original location) in order to obtain partial results referred to as *local knowledge models*. Then the local models are gathered and combined together to obtain a global model.

Also in this approach, for the local computations, it is possible to reuse sequential DM algorithms without any modification. The problem here is to identify a method to combine the partial results coming from the local

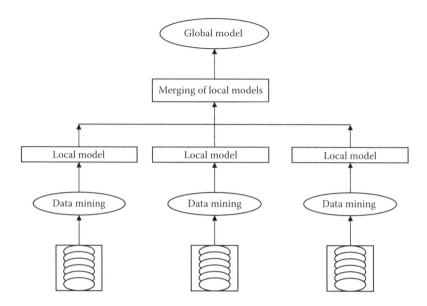

FIGURE 4.2   The model-driven approach for DDM systems.

models. Different techniques can be adopted, based on voting strategies or collective operations [3]. For example, multiagent systems may apply met-alearning to combine partial results of distributed local classifiers [4]. The drawback of the model-driven approach is that it is not always possible to obtain an exact final result. The Papyrus system [5] is a DDM system that implements a cost-based heuristic to adopt either the data-driven or the model-driven approach.

4.2.2.3 An Architecture-Driven Approach

In order to be able to control the performance of a DDM system, it is necessary to introduce an additional layer between data and computation. As shown in Figure 4.3, before starting the distributed computation, the data are moved to different sites with respect to their original location. Here a layer is introduced which forms the communication layer among the local DM computations, so that the global knowledge model is built during the local computation. In this approach, the focus is on optimal use of resources, and hence it is referred to as *architecture driven*. Figure 4.3 represents the architecture-driven approach of the DDM systems.

A notable example of such a system is the optimal data and model partitions (OPT-DMP) framework [6], which aims at balancing performance and accuracy in DDM applications.

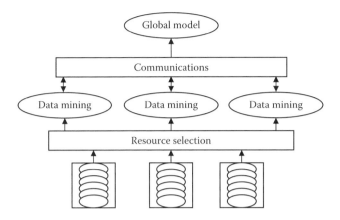

FIGURE 4.3   The architecture-driven approach for DDM systems.

4.2.3 State of the Art

Sequential DM, although still under development, is currently being applied in many application domains, and it can be considered an industry-ready technology, ready and mature enough for delivery at industrial standards.

4.2.3.1 Parallel and Distributed DM Algorithms

Many sequential clustering algorithms have been parallelized, for example parallel implementation [7, 8] of the popular K-means algorithm [9]. Talia et al. have parallelized the unsupervised classification algorithm "Auto class" [10].

Distributed classifiers have been studied [4] and their ability to operate on either homogeneous or heterogeneous data is observed. They typically produce local models that are refined after exchanging summaries of data with the other classifiers.

A different approach is adopted in the collective data-mining framework [11], where instead of combining partial models coming from local analysis, globally significant pieces of information are extracted from local sites.

In distributed association rule mining algorithms, the following major approaches were found. One is count distribution, according to which the transaction database is statically partitioned among the processing nodes, while the candidate set $-C_k$ is replicated. In each iteration, every node counts the occurrences of candidate item sets within the local database partition. At the end of the counting phase, the replicated counters are aggregated, and every node builds the same set of frequent item sets F_k.

On the basis of the global knowledge of F_k, candidate set C_{k+1} for the next iteration is then built. Internode communication is minimized at the price of carrying out redundant computations in parallel. As another example, data distribution attempts to utilize the aggregate main memory of the whole parallel system. Not only the transaction database but also the candidate set C_k are partitioned in order to permit both kinds of partitions to fit into the main memory of each node. Processing nodes are arranged in a logical ring topology to exchange database partitions, since every node has to count the occurrences of its own candidate item sets within the transactions of the whole database. Once all database partitions have been processed by each node, every node identifies the locally frequent item sets and broadcasts them to all the other nodes in order to allow them to build the same set C_{k+1}. This approach clearly maximizes the use of node aggregate memory, but requires a very high communication bandwidth to transfer the whole data set through the ring at each iteration.

A complete architecture for large-scale distributed data mining is the K-Grid. Built on top of standard Grid middleware, like the Globus Toolkit [12], it defines services and components specific to the DM and KDD domains.

4.2.4 Research Directions

DDM technology needs to be further studied and enhanced in order to reach industry standards. Hence there is the need for new distributed algorithms. It is necessary to develop new strategies for DM computations. One important feature that distributed DM algorithms should have is the ability to balance accuracy and performance. A large fraction of new applications deal with streaming data that often require real-time execution. There is therefore the need for online algorithms to be able to cope with the ever-changing flow of input information.

An important issue in recent years is privacy-preserving data mining [13, 14]. In distributed environments, it might not be possible to access some data onto which privacy constraints hold. Distributed algorithms should be able to handle privacy issues. An emerging technology includes data mining involving web services.

4.2.5 Scheduling DM Tasks on Distributed Platforms

The need for high performance makes large-scale distributed environments, like the Grid [15], a suitable environment for DDM. Grids provide coordinated resource sharing, collaborative processing, and high-performance computing platforms. Since DDM applications are typically data intensive, one of the main requirements of such a DDM Grid environment is the efficient management of storage and communication resources.

4.2.6 Data and the K-Grid

Data-intensive applications [16] consist of a data management architecture based on storage systems and metadata management services. The data grid framework consists of data that originate from a large variety of sources. A specialized grid infrastructure, the K-Grid, is discussed in this section. The K-Grid architecture is divided into two layers: the core K-Grid layer and the high-level K-Grid services layer (see Figure 4.4). The first refers to services directly implemented on top of generic grid services; the latter refers to services used to describe, develop, and execute parallel and distributed knowledge discovery (PDKD) computations on the K-Grid. Moreover, the layer offers services to store and analyze the discovered knowledge.

The KDS extends the basic Globus Metacomputer Directory Service (MDS) [17], and is responsible for maintaining a description of all the data and tools used in the K-Grid. The metadata managed by the KDS are represented through XML documents stored in the Knowledge Metadata Repository (KMR). Metadata consist of data source characteristics, data management tools, data-mining tools, mined data, and data visualization tools.

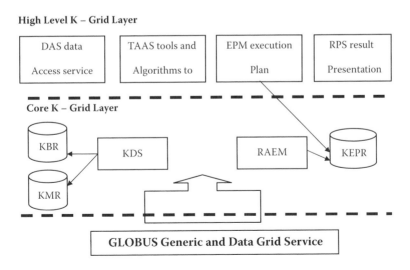

FIGURE 4.4   General scheme of the K-Grid software architecture.

The *resource allocation and execution management* (RAEM) service provides a specialized broker of Grid resources for DDM computations. The broker takes allocation and scheduling decisions, and builds the execution plan.

4.2.7 The Knowledge Grid Scheduler (KGS)

K-Grid services can be used to construct complex problem-solving environments. The K-Grid can perform an initial clustering on a given data set in order to extract groups of homogeneous records, and then look for association rules within each cluster. An example of such a system is the VEGA visual tool [18] for composing and executing DM applications on the K-Grid.

A general DM task on the K-Grid can therefore be described as a directed acyclic graph (DAG) whose nodes are the DM algorithms. The links represent data dependencies among the components.

In this scenario, one important service of the K-Grid is the mapping of task requests onto physical resources. The user will in fact have a transparent view of the system and possibly little or no knowledge of the physical resources where the computations will be executed. The user may not know where the data actually reside.

4.2.8 Requirements of the KGS

Many efforts have already been made for scheduling distributed jobs, such as Nimrod/G [19], Condor [20], and AppLeS [21]. The first phase is devoted to resource discovery. During this phase, a set of candidate machines where the application can be executed is built. This set is obtained by filtering the machines where the user has enough privileges to access and at the same time satisfy some minimum requirements, expressed by the user. In the second phase, one specific resource is selected among the ones determined in the previous phase. This choice is performed based on information about the system status for, as examples, machine loads, network traffic, and, again, possible user requirements in terms of the execution deadline or a limited budget. Finally, the third phase is devoted to actual job execution, from reservation to completion, through submission and monitoring.

4.2.9 Design of the KGS

The design of the KGS is described in this section. A model for the resources of the K-Grid, depicted in Figure 4.4, is composed of a set of hosts, on

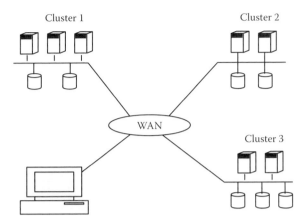

FIGURE 4.5   Physical resources in the K-Grid

which the DM tasks are executed. Figure 4.5 shows a structured way of forming a DAG. DM components correspond to a particular algorithm to be executed on a given data set. We can describe each DM component Λ with the three parameters *A*, *D*, and *P*:

$$\Lambda = (A,D,\{P\})$$

where *A* is the data-mining algorithm, *D* is the input data set, and {*P*} is the set of algorithm parameters. For example, if *A* corresponds to *association mining*, then {*P*} could be the minimum confidence for a discovered rule to be meaningful. It is important to notice that *A* does not refer to a specific implementation. Figure 4.5 describes the physical resources of the K-Grid.

In Figure 4.6, the original DM task on the left-hand side is composed of an application of a clustering algorithm on a certain data set, and then by the application of an algorithm for association mining on each cluster found. Finally, all the results are gathered for visualization. In the mapping reported on the right-hand side of Figure 4.6, a node is added on the top of the graph, which corresponds to the initial determination of the input data set. Figure 4.6 shows the case where one of the association rules' mining is performed on a parallel architecture. In this way, starting from a semantic DAG, a physical representation of DAG is defined (see Figure 4.7). Later, all the components are mapped onto actual physical resources.

This process is repeated for all the DAGs that arrive at the scheduler.

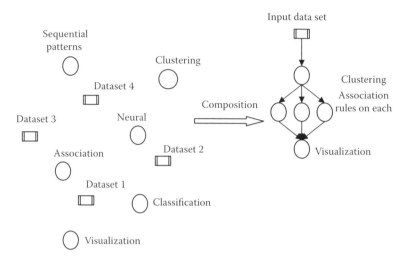

FIGURE 4.6 Composition of a DM DAG in terms of basic building blocks: data sets and algorithms.

4.2.10 An Architectural Model for a K-Grid

The K-Grid architecture is composed of a set of clusters. Every host has a storage unit and contains a subset of the data sets available in the system. A data set in general may be replicated at different sites. Hosts are connected by fast links in local area networks (LANs), thereby forming clusters. Slower links of a geographical wide area network (WAN) connect the clusters. The scheduler provides a front-end view to the users, and a coherent view to the system. From the front end, users can choose a data set and a DM task to be performed on the data set, among a list of possibilities.

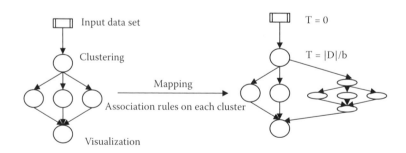

FIGURE 4.7 Mapping a DAG onto physical resources permits one to obtain cost measures for the DM task associated with the DAG.

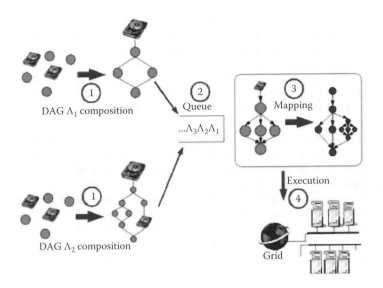

FIGURE 4.8 Complete K-Grid architecture.

A user request generates a job submission to the scheduler, which in turn will locate the data. The scheduler will choose the best algorithm implementation and will assign the task to a specific host or set of hosts. At this level, it is assumed that hosts are all locally controlled by batch queues that allow only one job to be executed at the time. Figure 4.8 represents the complete K-Grid architecture.

LAN and WAN links are described by a function of the number of the communications taking place. For WAN links transmitting big data sets, one can assume a constant bandwidth up to a given number of connections; after that, the effective measured bandwidth starts decreasing. If $f\,i,j$ is the function describing the effective measured bandwidth of the link between nodes i and j, and n is the number of files being transferred between i and j, choose f as

$$f\,i,j(n) = \begin{cases} b & if\ n \le \hat{n} \\ b\hat{n}/n & if\ n > \hat{n} \end{cases}$$

where b is the effective bandwidth and n is the maximum number of connections that preserve the service time. Figure 4.9 represents the simplified DAG-based description model.

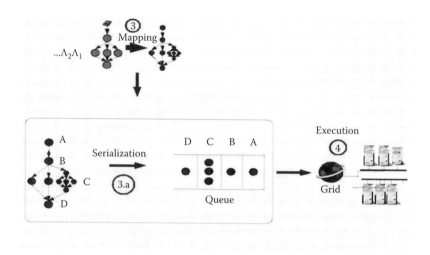

FIGURE 4.9   Simplified DAG-based description model.

4.3 CONTEXT-BASED SENSING, DATA MINING, AND ITS APPLICATIONS

Data mining, which aims at extracting interesting information from large collections of data, has been widely used as an active decision-making tool. Real-world applications of data mining require dynamic and resilient models that are aware of a wide variety of diverse and unpredictable contexts. Contexts consist of circumstantial aspects of the users and domains that may affect the data-mining process. The underlying motivation (i.e., mining data sets in the presence of context factors) may improve performance and efficacy of data mining by identifying the factors, that are not easily detectable with typical data-mining techniques. In a context-sensing data-mining framework, the context can be represented in a particular aspect and can be automatically captured during data mining. The different behaviors and functionalities of context-sensing data mining can dynamically generate information in dynamic, uncertain, and distributed applications.

4.3.1 Applications of Contextual Data Mining

Real-world applications are laden with huge amounts of data and encompass entities that evolve over time. However, this data-rich environment does not guarantee an information-rich environment. Due to the dynamic nature of environments, data must be interpreted differently depending upon the situation (context). For instance, the meaning of a cold patient's high fever might be different from the fever of a pneumonia patient. Context is a powerful concept. It can be helpful in human–computer interaction, which is running mostly via explicit contexts of communication (e.g., user query input). Implicit context factors (e.g., physical environmental conditions, location, and time) are normally ignored by the computer due to the absence of a knowledge base or appropriate model. However, not much work has been reported on building data-mining frameworks based on context sensitivity leading to useful and accurate information extraction.

Finding the influence of the context on the final decision made by the decision maker is an important task because it helps to find and model typical scenarios of interaction between users and to reveal typical situations within large amounts of raw data.

Analysis of different kinds of decisions is one of the areas of data mining for business processes. The goal of context mining is to find dependencies and associations, explaining under what circumstances a certain activity is to be selected rather than another one. For instance, decision mining aims at detecting data dependencies that affect the routing of a case in the event log of the business process execution. Context sense mining covers a wide range of problems. Estimation of data quality and interpretation of their semantics comprise one of the major tasks of context mining. It requires the following interpretation of data: whether it is relevant, what it actually means, in what units it is measured, and so on. Classification of different contexts is also one of the important tasks for context mining.

4.4 EXAMPLE: THE COASTAL RESTORATION DATA GRID AND HURRICANE KATRINA

A successful restoration of the Gulf Coast and the prevention of possible future disasters depend highly on careful analysis of the satellite and sensor data collected over long periods of time. Satellite cameras over the Gulf Coast periodically take high-resolution images, and sensors scattered across the Gulf capture continuous status data from the coast. The amount of coastal data produced is measured in hundreds of terabytes per year, and this amount continues to increase at an exponential rate. The current computing infrastructure is not yet able to store and process all of the data generated in a timely manner, and scientists have to disregard some of the data without even having an opportunity to analyze them. The development of contextual models for the selection of important data can change and mitigate this fact by pointing to key data and the environmental contextual data that should be selected for analysis. Currently such capabilities do not exist in a comprehensive, broad, and distributed sense.

4.5 THE POWER OF INFORMATION MINING IN CONTEXTUAL COMPUTING

Hurricanes, coastal erosion, drought, cyclones, and typhoons are major threats to human lives and infrastructure on a larger scale. The scale and complexity of these threat-warning systems present fundamentally new research challenges in distributed systems paradigms. Innovative applications of grid computing and distributed sensor networks (DSNs) are increasing the pace and scope of research activity as these technologies move closer to becoming a reality for many real-time data-intensive civilian or defense applications. The power of data mining is one of the underlying themes toward a more common vision to improve the core significant research progress around the world. Data mining, however, has the problem of selecting the correct set of data to do analysis on. In this sense, contextual-processing models can be beneficial in the selection of the key data sets for more powerful data mining. This new paradigm might be called *contextual data mining*, where existing methods are applied to contextually selected sets of data. This can play a leading role in research-oriented applications of DSNs toward real-time contextually based processing of massive amounts of data.

Natural disasters' effects on ecological systems have highlighted the urgent need for interdisciplinary initiatives into research collaborations among all the areas of information systems, such as those found in global contextual-processing systems architectures. Contextually based models for computational modeling, simulation analysis, and the utilization of large-scale cyber-information technology enabled by Internet access could provide external facets into further research into this natural phenomenon.

4.6 ENABLING LARGE-SCALE DATA ANALYSIS

The high computational and storage requirements of coastal data necessitate the use of distributed resources owned by collaborating institutions to meet the storage and analysis demands. In such a distributed computing environment, data are no longer locally accessible to the coastal analysis and simulation applications, and have to be remotely retrieved and processed, which brings extra challenges. Additionally, available data resources and computational resources might be unknown to consumers of such information, which points in the direction of hyperdistribution and discovery of computational resources and data, as will be discussed in the following chapters.

Produced data need to be preprocessed to allow one to archive only the useful data and discard the rest. Archived data need to be cataloged based on contextual classes and concepts presented in Chapter 2 to enable easy access to the desired segments, and related data must be managed in a fashion that will preserve contextual relationships. Multiple copies of high-priority information could be generated and stored at different physical locations to increase reliability and also enable easier retrieval by the researchers from different geographical locations. This architecture demands *contextual maps* of the knowledge space and distribution of information. In the future, contextual research in this area could move data to computation sites for processing based on contextual processing, contextual relationships, and contextual hyperdistribution mechanisms. This end-to-end process should be flat in control and not hierarchical. It is a space where research will develop contextually based methods to manage data and coordinate processes to enable reliable and efficient processing and analysis of the large amounts of coastal data based on contextual modeling. Current methods for doing this do not have such capabilities, especially when the information location or existence of information can be unknown or underdetermined. Additionally, the overwhelming amounts of data that could be processed make it hard, at present, to select what is key and what is not because contextual models of relationship have not been applied to them.

4.7 EXAMPLE: ACCESSING REAL-TIME INFORMATION—SENSOR GRIDS

Most calamities strike within a very short period but have a huge impact. Calamities like tsunamis, tropical storms, hurricanes, and typhoons have claimed many lives and caused damage costing many billions of dollars. A real-time sensor-based contextually modeled disaster management system for hurricane and tsunami detection would enable a faster transfer of critical data around the globe and sound warnings even though the center of activity may be far away from a particular coastal area, giving precious additional time to evacuate these regions and save many lives. The system can also be used to collect information about water level in the seas and seismic activities, and hence will foster oceanographic research in general. Many problems in the aftermath of Hurricane Katrina were caused by a lack of information flow between different federal, state, and city agencies, so solutions to the problem of contextual data and knowledge integration will be very useful in the rebuilding effort and the prevention of future disasters.

This chapter keys in on the notions that current data-mining methods, often working on large volumes of information collected from sensor networks, often have to store data at various sites for post analysis. It also highlights the need for better dissemination of critical information about natural disasters. Contextually based models can be developed that could be utilized for selection of initial data sets for data analysis, where the key principles in Chapter 2 could be applied to determine critical contextual relationships in data set generation. Contextual processing could then distribute information using the concepts and theories of hyperdistribution and contextually based information management proposed in the following chapters.

4.8 RESEARCH DIRECTIONS FOR FUSION AND DATA MINING IN CONTEXTUAL PROCESSING

Complex situations like prediction of natural disasters involve a great number of different heterogeneous data, sometimes multidimensional, which have to be combined in order to succeed. Often it is necessary to use external sources to get the required information. Such actions require intensive information exchange in order to achieve the necessary level of situational awareness, the creation of ad hoc action plans, and continuously updated information. Centralized control is not always possible due to probable damages in local infrastructure and uncertain cooperation of participating teams. A possible solution for this is the organization of decentralized self-organizing coalitions consisting of the operation's participants. Some of the problems to study include providing for semantic interoperability between the participants and defining the standards and protocols to be used. For this purpose, such technologies as situation management for natural disasters, profiling, and intelligent agents can be used. Standards of information exchange (e.g., web service standards), negotiation protocols, decision-making rules, and so on can be used for information knowledge exchange.

The focus of this chapter is on data mining and its distributed representation with context-sensing applications. More research related to the area of data mining in context-sensing intelligent applications is necessary. An approach for natural disaster applications like tsunamis, tornados, hurricanes, avalanches, earthquakes, floods, droughts, fire, volcanic eruptions, landslides, and so on which carries lots of auxiliary data needs to be developed. In the user profile, context sensing can be associated with each decision-making process. These data can be text, images, or any other multidimensional element describing the actual situation, based on the natural disaster applications which are available in geographic information systems. A set of alternative solutions and the final decision can be proposed with the help of context-based sensitive mining for revealing the predictions which will influence the requirements of the final results. Lots of accumulated data are available from information systems on natural disasters that have already taken place, which can be context sensed for taking care as a precautionary measures against future disasters.

The next chapter will present a key notion in contextual processing, hyperdistribution. Simply put, in a global contextual system, there will be consumers of information that will not know that producers of information of interest exist. Thus, Chapter 5 looks at the process of connecting consumers and producers of information who lack a priori knowledge of each other's existence.

REFERENCES

1. Parkinson, C. L., and R. Greenstonen (eds.). (2000). *EOS data products handbook*. Greenbelt, MD: NASA Goddard Space Flight Center.
2. Chervenak, A., I. Foster, C. Kesselman, C. Salisbury, and S. Tuecke. (2001). "The data grid: Toward an architecture for the distributed management and analysis of large scientific datasets." *Journal of Network and Computer Applications* (23): 187–200.
3. Kargupta, H., W. S. K. Huang, and E. Johnson. (2001). "Distributed clustering using collective principal components analysis." *Knowledge and Information Systems Journal* 3 (4): 422–448.
4. Prodromidis, A. L., P. K. Chan, and S. J. Stolfo. (2000). "Meta-learning in distributed data mining systems: Issues and approaches." In *Advances in distributed and parallel knowledge discovery*. Cambridge, MA: AAAI/ MIT Press.
5. Grossman, R., S. Bailey, B. Mali Ramau, and A. Turinsky. (2000). "The preliminary design of papyrus: A system for high performance, distributed data mining over clusters." In *Advances in distributed and parallel knowledge discovery*. Cambridge, MA: AAAI/MIT Press.
6. Grossman, R., and A. Turinsky. (2000). "A framework for finding distributed data mining strategies that are intermediate between centralized strategies and in-place strategies." Paper presented at the *KDD Workshop on Distributed Data Mining*, Boston, December.
7. Baraglia, R., D. Laforenza, S. Orlando, P. Palmerini, and R. Perego. (2000). "Implementation issues in the design of I/O intensive data mining applications on clusters of workstations." In *Proceedings of the 3rd Workshop on High Performance Data Mining (IPDPS-2000)*, Cancun, Mexico, pp. 350–357, Berlin: Springer LNCS.

8. Holland, J. H. (1975). *Adaptation in natural and artificial systems.* Ann Arbor: University of Michigan Press.

9. Jain, A. K., and R. C. Dubes. (1988). *Algorithms for clustering data.* Englewood Cliffs, NJ: Prentice Hall.

10. Pizzuti, C., and D. Talia. (2003). "Scalable parallel clustering for mining large data sets." *IEEE Transactions on Knowledge and Data Engineering* 15 (3): 629–641.

11. Park, B., and H. Kargupta. (2002). "Distributed data mining: Algorithms systems and applications." In N. Ye (ed.), *Data mining handbook.* Hillsdale, NJ: Erlbaum.

12. Foster, I., and C. Kesselman. (1997). "Globus: A metacomputing infrastructure toolkit." *International Journal of Supercomputer Applications* 2 (11): 115–128.

13. Lindell, Y., and B. Pinkas. (2000). "Privacy preserving data mining." *Lecture Notes in Computer Science* 1880: 36.

14. Agrawal, R., and R. Srikant. (2000). "Privacy-preserving data mining." In *Proceedings of the ACM SIGMOD Conference on Management of Data*, pp. 439–450. New York: ACM Press.

15. Foster, I., and C. Kesselman. (1999). *The Grid: Blueprint for a future infrastructure.* San Francisco: Morgan Kaufman.

16. Fitzgerald, S., I. Foster, C. Kesselman, G. von Laszewski, W. Smith, and S. Tuecke. (1997). "A directory service for configuring high-performance distributed computations." In *Proceedings of the 6th IEEE Symposium on High Performance Distributed Computing*, pp. 365–375. Piscataway, NJ: IEEE Computer Society Press.

17. Cannataro, M., D. Talia, and P. Trunfio. (2002). "Design of distributed data mining applications on the knowledge grid." In *Proceedings of the NSF Workshop on Next Generation Data Mining*, pp. 191–195, November. Baltimore: NSF.

18. Abramson, D., J. Giddy, I. Foster, and L. Kotler. (2000). "High performance parametric modeling with Nimrod/G: Killer application for the global grid." Paper presented at the *International Parallel and Distributed Processing Symposium*, Cancun, Mexico, May.

19. Epema, D. H. J., M. Livny, R. van Dantzig, X. Evers, and J. Pruyne. (1996). "A worldwide flock of condors: Load sharing among workstation clusters." *Journal on Future Generations of Computer Systems* 12 (2): 53–65.

20. Berman, F., R. Wolski, S. Figueira, J. Schopf, and G. Shao. (1996). "Application level scheduling on distributed heterogeneous networks." In *Proceedings of the 1996 ACM/IEEE Conference on Supercomputing.* Pittsburgh, PA: ACM.

CHAPTER 5

Hyperdistribution of Contextual Information

THEME

In previous chapters the concepts and constraints of a context model were developed. Reasoning methods about the similarity of contexts were discussed. Additionally, the notion that contexts operate in a fashion similar to that of sensor networks in the production of data was discussed through the concept of data fusion. Unlike a sensor network, in the contextual model there are actually two types of sensor networks: one producing data, and one consuming data and performing secondary processing and production. Thus, in the second network, information is consumed, context processing is performed, and this then becomes a derivation of the original data-producing network that can produce data of its own and control the contextual dissemination of information.

This chapter will explore the issues of distributing contextual information via hyperdistribution, a world where central hierarchical control does not exist and authority is based on a flat model—where there are no "super"-peers—such as that found in a peer-to-peer (P2P) network. We next examine and build a model of constraints on how such a system might operate. Methods and vehicles that support hyperdistribution within the constraints of how it must operate are then presented and discussed. Finally, advanced topics that might be addressed in such a model are introduced. These include the aggregation of consumer sensors into supernodes and methods that assist in the modeling of such systems and infrastructures. Several formal modeling tools are discussed that provide the structure necessary to model the hyperdistribution of contextual information as introduced in this chapter. Utilizing such tools to model a hyperdistribution system would provide methods to validate the functionality of the system and to furthermore quantify its performance, efficiency, and scalability. The eventual goal of such a discussion would be the creation of a model where information flows to where it is needed and can be edited and disseminated by interested parties based on trust relationships.

5.1 INTRODUCTION TO DATA DISSEMINATION AND DISCOVERY

There is a wealth of data that resides throughout the world in many forms; for example, records in relational database tables, flat file sets containing various kinds of output generated by sensors, shape files capturing the details about a geographical location, and websites containing everything from the latest political news to detailed instructions outlining a cheap and clever method to build a light-following robot. There is also a rather large amount of redundant data, much of which we know nothing about simply due to its magnitude. Much of these data can be used to eventually provide us with information, that is, something meaningful in some way to us.

In order to obtain information (from the data), however, it is often necessary to gather the surrounding context. Take, for example, a person yelling, "Fire!" What exactly is meant by that word? Perhaps it is nothing but signaling the end of the countdown to a launch, or perhaps it implies the excitement of music being played around a mesmerizing bonfire at a family reunion. In the extreme case (perhaps the first that is thought of), it may mean that there is indeed a fire in our currently occupied building and that we need to get out as quickly as possible! The context, then, provides the necessary surrounding information that is required in order to accurately process the data and potentially perform some sort of action.

Context data are furthermore being generated at many places around the planet, and many people might want to acquire and process such data. For the purpose of this chapter, we will call these "context generators" *producers*. They gather and process context data. At many other places, there are people who may want certain types of information but are unaware that it is being created by the producers. We will call these *consumers*. They may possess data already but are unaware of what to do with it because they lack the necessary surrounding context. Provided with the context, they can then perform some action.

The hierarchical system currently employed by many entities around the globe has many points of failure. There is a centralized control and, as a result, a reliance on a limited number of servers or *centers* that are used to route and disseminate information. From there, information is filtered down until it reaches its destination. Take the central servers down and you are no longer distributing information, which could potentially result in havoc. A distributed method that does not rely on any central servers or centralized control, such as P2P networks, provides a more reliable method to disseminate information to where it is needed. However, the problem of accurately disseminating contextual information to needy consumers in a nonhierarchical system must be addressed.

The questions we are attempting to answer in this chapter essentially involve finding a way to connect the producers and consumers. Intrinsically, there are a number of challenges related to the hyperdistribution of contextual information. For example, how is contextual information generated? How do the producers of information locate consumers efficiently in a world of millions of computers? Once a consumer is located, how is the contextual information routed efficiently? How does all of this occur without a centralized control? In this chapter, we provide our view of hyperdistribution within the contextual domain and our solution to connecting the producers and consumers, and discuss other advanced topics that can assist us in this task.

5.2 DEFINING HYPERDISTRIBUTION

At its most basic, hyperdistribution involves disseminating contextual information efficiently from producer to consumer, that is, the union of producers and consumers. Ultimately, however, many questions must be answered; for example, how do consumers obtain the necessary context for the information they possess? How do producers locate *hungry* consumers? How is the appropriate context manufactured and provided? What do the consumers ultimately do with the context? To this end, we define *hyperdistribution* within the contextual domain to be a series of ordered processes, where information is offered based on several criteria:

- *Ontological interest.* Relationships among information are established via an ontology. This is used to direct the information relevance.

- *Contextual connection.* Information may be "connected" contextually, and this may ultimately direct the flow of information.

- The *role* of producers and consumers in the "world."

- *Relationships* among producers and consumers. This is similar to the way we as humans relate to others; for example, if a producer *knows* a consumer, then they establish a relationship that may control how information is offered.

Definition: Hyperdistribution *is the process by which contextual informa-tion flows to where it is needed within the context of contextual processing. It is composed of four sequential processes that involve producers and con-sumers: discovery, validation, activation, and dissemination of contextual information.*

In detail, the four sequential processes that comprise hyperdistribution are as follows:

- *Discovery:* The process by which a producer discovers a consumer that might be interested in a type of thematic information. Discovery of all consumers of information worldwide works mathematically by the principle of *degrees of separation*, as in social networks. It also works by virtue of complexity theory, where simple organism asso-ciations create wide networks. The basis for discovery is the *knows relationship*, which exhibits a mathematical transitive property; for example, $a = b, b = c \rightarrow a = c$. Discovery is based on a service that examines a consumer's *interest ontology*. This ontology describes metaclasses of interest to the consumer and, for security purposes, may be published on a local server where the discovery service exam-ines and then sends an encrypted message invitation and certificate to the consumer offering contextual information.

- *Validation*: The union of producer and consumer occurs on the fol-lowing conditions:

  1. An interest ontology for a consumer has been discovered.

  2. An encrypted invitation has been sent to the interested consumer.

  3. The consumer has accepted the offer.

4. The connection context has been validated and consists of the following elements:

- *Trust level*: How much do we trust the consumer?

- *Knows relationship*: What basic attribute describes our knowledge of the consumer?

- *Temporal germanity*: How long have we known the consumer, and does this affect the *knows relationship* (e.g., less than one day, one day, from now on, always, never, sporadically, or continuously)?

- *Spatial germanity*: From where do we know the consumer, and does this affect the *knows relationship* (e.g., location)?

- *Impact*: Does our relationship have an impact on the connection context?

- *Similarity* to other producers.

- *Confidence* in *knows relations*: How confident are we in our relationship with the consumer?

- *Plausibility*: Does the consumer appear to require this information?

- *Is-a relationship*: Can we validate the consumer?

- *Has-a relationship*: Does the consumer have a need to know, an interest in the information, a right to know, or the like?

- *Activation*: The consumer connection context has been validated and a secure connection to the producer is established. During this process, several actions must be performed:

1. Exchange of duration of the connection

2. Time-to-live connection settings (e.g., n time, always, or sporadic disconnect)

3. Activation of a method on a web service

4. Agent download

- *Dissemination*: When the "real" work is done and contextual information flows from producer to consumer. In this process, the consumer performs two tasks:

  1. Accepts contextual information for connection parameters, for example time and dissemination level (whether classified, broadcast, or unspecified).

  2. May be promoted to producer at this point based on numerous factors such as role, *knows relationship*, and reliability. If this occurs, the consumer iterates to process discovery.

Our vision in hyperdistributing contextual information lies in the application of a multiagent system (MAS) whereby hyperdistribution becomes a cooperative activity between intelligent agents, some of which are producers, some of which are consumers, and some of which are both (consumers may become producers of the contextual information they have received). At the core of the MAS are two main components:

- *Intelligent agents*: Representing producers and consumers, they autonomously facilitate the hyperdistribution of contextual information.

- *Agencies*: These provide agents with an interface to services (e.g., computational resources, and data) and a method of meeting other agents in a group setting. Agencies also act as sandboxes, restricting access to the physical hardware and resources. This concept offers a way to group agents cooperatively.

Agents are autonomous and mobile (i.e., they can migrate from agency to agency) but have a *home* agency that serves as their primary location. The main goal of a consumer is to go around gathering information and its surrounding context so that it can achieve its goals. As a result, it constantly migrates from agency to agency. Producers go around perpetually gathering information and computing contexts.

Some agencies are also called *markets* in that they serve as a place for consumers to advertise their needs for contextual information and producers to provide those needs. In the case that a consumer must migrate to another agency, markets provide *bulletin boards* where a consumer may leave a message requesting contextual information. Producers that arrive at the agency at a later time have access to this bulletin board. In this way, consumers can be provided with the contextual information they require even while they are migrating to other agencies throughout the network. Producers and consumers can also make contact separately (i.e., without the need for an agency). Markets facilitate the dissemination of information, thereby providing a natural flow of information between producers and consumers.

To successfully compute the context surrounding some piece of information, the domain of knowledge must be restricted to something fairly manageable. What concepts are consumers interested in? Suppose a consumer obtains the information *coffee*. Alone, it may mean nothing. Suppose the context *breakfast* were provided. The assumption that *breakfast* is understood by the consumer is made. As a result, context must have some sort of predefined ontological structure. We begin with a simple domain and make several assumptions (e.g., everyone speaks English). From a fixed ontology, we will extract the context necessary to surround some piece of information in order for a consumer to capably perform some action. Each agent in the MAS will utilize the ontology in his or her reasoning.

In the worst case, context can be everything in the universe. The strategy, then, is to constrain the context by specifying an agent model. We initially specify a set of rules that defines an agent. Each rule requires three components:

1. *Information*: Cannot be adequately interpreted without context

2. *Context*: Allows the information to be correctly interpreted

3. *Action*: Can be performed by a rule if the context provided by the producer clarifies the information the consumer possesses

We suppose the agent has some information it needs context for in order to determine if it must perform some action. The context is computed by and obtained from the producers. In order for a producer to capably compute the context, it must know something about the consumer it is preparing the context for. For this, we propose a situation model that provides the consumer's situational perspective to the producer. This will assist the producer in computing the proper context by using the ontology. The situation may, for example, provide information relevant to identity, locality, and time. The situation model illustrated in Figure 5.1 represents the consumer's situational perspective to the producer. A consumer's actions depend on how it interprets the information it has via the context it is provided by the producer; this context may further depend on the consumer's

FIGURE 5.1   Situation model.

identity, locality, and policies, among other things. There may be numerous possible contexts that surround a piece of information. Selecting the appropriate one may depend strictly on the consumer's locality. This is particularly true when dealing with natural disasters; for example, the fact that a tsunami is occurring on the opposite end of the globe may not be a concern to a particular consumer.

It is important to also consider the result of an agent's actions. In essence, an action is complete only if a result is produced. However, it is entirely acceptable to have a result that is null. Being mindful of action results allows us to say something meaningful about contextual processing. Furthermore, results may be utilized at a later time to assist in making decisions. One may think of the process of obtaining and keeping results for the future as a way of increasing a level of experience. Producers and consumers may then be rated or ranked on such a level if desired.

The advantage of a multiagent approach is that the context computation can be performed on nodes that have the necessary data and computing power. The result—the context—is then the only thing that needs to be transmitted by the producer.

In hyperdistribution, context depends upon some object x and for whom we are building the context (the agent or consumer) A. Note that context cannot exist without producers and consumers since context is dynamically computed for specific consumers. G represents the universe of global information. A represents an agent (consumer). Given some information x, we wish to compute the context $C(x,A)$. Note that $C(x,A)$ is a subset of G (the subset that is relevant to the producer or consumer). Since the context is specific to a consumer's need, we specify a set of rules composed of state and behavior that each agent possesses in order to broadcast its need. A rule R is given as follows: $R = \langle R_1, R_2, \ldots, R_n \rangle$. Each rule uses x and $C(x,A)$ in order to determine if it will perform some action. For example, suppose $R = \langle R_1 \rangle; R_1 = x > 3 \wedge y < 5 \rightarrow$ *send an alert*. The consumer possesses the information x. The context computed by the producer will provide y. Producers will examine a consumer's rule base to identify resources used in conditions and actions. For example, suppose x is rain and $R_1 =$ *If it rains, then use an umbrella*. If A already has an umbrella, then no context is necessary. But if $R_1 =$ *If it rains, there will be a flood*, the flood information needs to be received by the producer. In general, there are three rule cases:

No context: This rule does not require a context; for example, R_1 : $cond(x) \rightarrow action\ a_1\ R_1$.

No information: This rule does not require any information (only a context); for example, $R_2 : y$ from $context \rightarrow action\ a_2$.

Context and information: This rule requires both information and a context; for example, $R_2 : cond(x) \wedge y$ from $context \rightarrow action\ a_3$.

Rule R_1 is strictly a local activity in that it requires no context. Rules R_2, R_3 are context-dependent activities. Several conditions may affect the ultimate result and if the consumer is able to perform some action. It is the producer's responsibility to compute the context. However, if it cannot determine the context, the consumer must then decide if it should continue without the context or randomly determine a context.

Different agents may require a different context for the same information. For example, suppose x is *fire* and A is *a child*. The context surrounding *fire* may be very different for a child than it would be for an adult. A child may always run away from a fire, whereas an adult may put it out with a water hose if the context surrounding fire triggers that action.

If an agent is capable of planning, then the context is temporal and has a time parameter. For example, the length of time it has been raining makes a difference in the case of a flood zone. If it has been raining but there is no flood, the consumer may make a mark in its calendar (schedule a task) that periodically checks to see if it is still raining. If it keeps raining, then the consumer will need to perform some action. The implication that an agent can plan introduces a complexity that must be addressed with respect to context computation by means of fuzzy logic or other techniques.

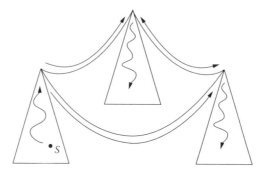

FIGURE 5.2   Context propagation in the traditional flat system (*s* is event source).

The traditional method of distributing information relies on a hierarchical structure that exhibits multiple weaknesses. In the typical case, there is a centralized control that creates a single point of failure for the entire system. In some cases, an extension of this involves the implementation of multiple primary *centers*—each of which can be a point of failure for a major part of the system—that communicate with each other and join prominent components of the system. If the connection between centers is broken, dissemination to a large part of the system is thereby halted. The overall structure takes on a pyramid-like form as shown in Figure 5.2, and if dissemination is disrupted at the top, the effects are felt on a grand scale and involve a large number of entities within the system. There is also a need to know of each member (e.g., by location), and relationships and policies must be established that aid in the dissemination of information.

The main advantage that hyperdistribution provides is a lack of centralized control. Exhibiting the characteristics of a flat system and akin to P2P networks, a failure in a single point only affects that entity; others are left unaffected by the failure. Furthermore, information flows only to those that need it as opposed to forced communication to major routing entities—as in hierarchical systems. A benefit is that each peer in the network provides necessary resources, thereby reducing the need for centralized pools of bandwidth, computing power, storage space, and so on. Our hyperdistribution model is intrinsically scalable in that as the population of producers and consumers grows, the total capacity of the system also increases. Hierarchical systems generally maintain a fixed set of resources available to all peers. Therefore, as the number of peers increases, the ability to manage them all decreases.

5.3 ISSUES IN HYPERDISTRIBUTION

Although in principle the hyperdistribution model offers scalability, robustness, and redundancy, it invariably introduces several issues that must be addressed:

1. How do producers *generate* contextual information?

2. How are consumers *discovered* in such a system?

3. How is the information *routed* appropriately?

5.3.1 Context Generation

In order to determine the relatedness of information, context generation requires an ontology. As a result, a user model for producers and consumers is required in order to make use of the ontology. In our agent model, consumers are composed of rules that may require context in addition to gathered information in order to determine if an appropriate action should be taken. Consumers are composed of two threads of control:

1. *Context building*: The process by which a consumer sifts through its rules for those that require contextual information and computes the context. In many cases, the context cannot be determined; this is where the producers come in. We assume that producers are in a better position to build context since they generally have more information.

2. *Query formulation*: If the context cannot be computed, then it must be requested. This process supplies producers with well-formulated queries so that they may provide the appropriate context when requested to do so.

Producers continuously build context. They are composed of two independent threads of control:

1. *Context creation*: The process of building a scenario model (context) utilizing an ontology.

2. *Context derivation*: By examining a consumer's rules and query—and using the ontology—the context is derived. It is important to note that the producer must identify the relevant part of the context to send to the consumer. The consumer must then dialogue with the producer to identify the relevance (i.e., the entire context may be too much information).

At any point in time, the producer computes its "world state," which is continuously updated so that when a consumer requests some context, it gets the latest information. It produces (builds) context in its complete form as much as possible. It must first construct the context and then distribute the context (using an MAS in our implementation). The producer must identify the relevant part of the context to send to the consumer—who must then communicate with the producer to identify the relevance. It may also be necessary to communicate the contextual level of abstraction to the consumer—ignoring unnecessary details—so that it may better interpret the context. For example, suppose the context of *fire* is *emergency*. There is no need to include a potentially large amount of information that resulted in this decision.

Computing the context requires the ontology. For example, suppose x is *fire*. The producer initially obtains the definition of *fire* in the ontology. It then examines the consumer's rules to see if any additional information is required in order to be able to interpret *fire* in the context. It may be that not all rules depend on the information *fire*; perhaps several depend on it, a few depend on the context surrounding it, and the remaining rules have nothing to do with it. As a result, there may be too much information for the producer to sift through; a strategy may be to use the ontology to act as a rule filter. It provides concepts related to *fire*. These concepts may then be used to determine which rules are necessary in order to compute the context.

Continuing with the example, suppose someone informs agent A that there is *fire*. A needs to act, but does not know the context in which *fire* has occurred. The context essentially behaves as extra information that further qualifies and describes *fire*. This issue begs the question: is this extra information always considered contextual? Consider a 911 emergency phone call. Yell "Fire!" into the phone, and the person on the other end will ask you the context. They will ask a series of questions in order to be able to appropriately provide a response. This constitutes a series of communications. In a sense, this is similar to an expert system. But what questions does one ask? A strategy might be to look at an agent's behavior rules and ask relevant questions based upon those rules. A may possess a rule that states, "If a fire is large and if it is in the forest, then run away." If A knows that there is a fire—but not where—the producer may be able to contribute this critical information by looking at A's rule base and determine if fires happen to occur in the forest frequently at this time of year. Furthermore, the producer may infer a fire in the forest if it is particularly dry this time of year, if it is summer, or the like.

One area of extension involves context computation in the absence of complete information. Context representation should ideally include some sort of fuzziness to represent uncertainty. Fuzzy logic allows us to generate the context $C(x,A)$ without having all of the information x. In the ideal case, a producer must have *general* knowledge (i.e., the producer is an expert). But rarely do we have an ideal case, so fuzziness and randomness come into the picture. Context inference rules may help to compute the context. Moreover, the entire context may be too much for the consumer. It may behoove us to produce an abstraction strategy such that a producer can communicate the level of abstraction along with the context.

5.3.2 Discovery of Consumers

The discovery of consumers requiring context poses an interesting problem. Ideally, consumers should *advertise* that they are hungry for certain types of contextual information, akin to a chirping baby bird that is hungry and waiting for food. Some consumers are already getting context information—the link has been established with the source. They are also listening for consumer *chirps* for contextual information in their domain and ultimately become suppliers to these consumers. In turn, they morph into linked consumers seeking other chirping consumers so that consumers in effect become ad hoc producers. Ideally, information should flow where it is needed—much like an amoeba.

5.3.3 Routing of Data and Contextual Information

Within the MAS as introduced above, interagent dissemination of information occurs concurrently as it is necessary. Both producers and consumers may indeed act as *routers* conveying contextual information to others in the system. One of the advantages of utilizing intelligent agents that behave as producers and consumers is that they can autonomously communicate with each other on behalf of each other. So an important context that one consumer happens to require may be provided by a producer known and accessible only through a series of other consumers and producers. Utilizing the trust model (i.e., *knows*), such relationships are easily established.

A series of agencies, called *markets*, provide the necessary supporting infrastructure for the consumers and producers to properly migrate from place to place and communicate.

Definition: A market *is a place where agents can meet and exchange contextual information. Physically, a market resides on a server or perhaps a group of servers. It provides (1) the mechanisms necessary to accept and authenticate migrating agents, (2) an interface to local resources (e.g., data and computational power), and (3) a system like a message board that allows agents to leave messages to and requests of other agents (i.e., a way to advertise the need for contextual information and to provide it).*

Relating to the infrastructure, markets provide a sandbox that restricts the agents in the system when localized on a particular machine. In this way, we distinguish them from other forms of mobile code, for example malicious worms and viruses which may have free and unlimited access to the machine. A producer may notice that a consumer is requiring some sort of contextual information that it happens to possess. The consumer leaves these messages on agencies as it traverses across the globe and may provide information identifying how to contact it with the needed context. It is certainly possible for agents to communicate directly through the network; but since the network is dynamic and a consumer's location will frequently change, such a method of communication may not always be possible. Utilizing the market infrastructure, consumers can be provided with required contextual information without necessarily knowing where to find the producer(s) that can provide it.

5.4 METHODS INFRASTRUCTURE, ALGORITHMS, AND AGENTS

In this section, we present the theory and reasoning for utilizing an MAS as the hyperdistribution model. Furthermore, we provide a brief overview of selected methodologies that assist in the modeling of hyperdistribution, which is useful to provide quantifiable validation and performance evaluation of the system.

5.4.1 Introduction

There is currently an immense push to migrate from a wired perspective to a wireless one. We see this with the growing trend of mobile device use. The traditional method of connecting to a network via cables is now being replaced with wireless network cards and routers, thus providing untethered network connectivity. Mobile computing is indeed becoming more prevalent; wireless networks are popping up everywhere.

Notwithstanding their usefulness and growing technological advancements, mobile devices do come with several disadvantages. They do not typically maintain continuous network connectivity in order to conserve battery life. Moreover, they usually have less computational power than their larger "desktop" counterparts. It is therefore quite common to distribute applications and services intended for users of mobile devices in order to address these and other concerns. The ideal hyperdistribution environment is well distributed.

The mobile agent paradigm has shown promise in addressing challenges inherent to distributed systems and has proven to be quite useful in addressing issues pertinent to computing on mobile devices. It is worth mentioning the stark difference between mobile *computing* as mentioned above and mobile *computation*. The latter embraces the idea of physically moving the computations themselves from one location to another. In this scenario, we might see a complex mathematical computation being initiated on a mobile device but performed elsewhere on a fixed computing resource—perhaps where the necessary data reside. The result is then delivered to the mobile device once computed. This is where mobile agents demonstrate their strength.

Mobile agents have numerous benefits; for example, they have been shown to significantly reduce bandwidth requirements and are quite tolerant of network faults [1]—both of which benefit users of mobile devices. Furthermore, they are well suited to carry out coordinated distributed tasks. They are being used in numerous ways: for example, in e-commerce [2]; network monitoring [3]; intrusion detection and protection [4]; data gathering, integration, and fusion [5]; resource discovery; and dynamic software deployment. In hyperdistribution, mobile agents function as producers and consumers that migrate from P2P and market to market, advertising their needs and providing contextual information when they can.

Web services are common and readily available, and uniquely provide an open standard for a secure distributed service-oriented architecture (SOA). Furthermore, accessibility to web services allows providers an expanded reach of their services, especially with the growing trend of mobile device use.

SOA is a model that provides the mechanisms necessary for the integration and cooperation of service-providing systems. It requires that participating systems expose the services they wish to provide via a common interface. These services are then invoked from a service provider to accomplish some desired task.

One approach to standardizing the mobile agent paradigm is to combine the flexibility of mobile agents with the availability of web services. By utilizing web services to facilitate agent migration, we can standardize the assimilation process, thereby making it less painful and involved to implement.

5.4.2 Intelligent Agents

The intelligent agent was born as a means to provide a solution to the dilemma of autonomously performing tasks for users. Strictly speaking, an agent is some sort of object that performs a task on behalf of a user. More specifically, Wooldridge and Jennings in [6] give the intelligent agent several unique characteristics:

- *Autonomy*: An agent is capable of performing tasks without requiring user intervention; furthermore, an agent may implement mobility aspects and be able to move from one location to another.

- *Social ability*: An agent is capable of communicating with other agents and/or other entities; it may adapt its behavior accordingly based on the communication activities it performs.

- *Reactivity*: An agent can react to changes in the environment or in events that occur within that environment.

- *Proactiveness*: An agent can not only react to changes and events but also induce changes and initiate patterns of behavior when necessary.

Intrinsic to the agent is the concept of an agency. Essentially, an agency is nothing more than a storehouse for agents and is generally software based. The agency may accept a group of agents by providing an interface that supports a set of tasks; for example, querying a relational database on the network. Often, agencies are utilized as a means to provide safe and secure interagent communication. *Markets* are a form of agency that assist in the discovery of producers, consumers, and services.

With respect to an agent, the concept of intelligence is one on which many in the discipline of computer science opine. In the field of artificial intelligence (AI), it generally means that an agent exhibits characteristics similar to humans, such as intention and motivation [6]. Concepts central to AI such as fuzzy logic, expert systems, and artificial neural networks have been used to implement the *intelligence* of an agent. It is important to note, however, that this intelligence is not a strict intelligence; that is to say, it is not human intelligence but rather computational intelligence [7]. An intelligent agent can be given a set of rules to follow, some of which may actually be to learn from experience. In turn, as the agent performs more tasks, it can become better at performing them.

Recently, intelligent agents have been used to aid in gathering data on the web. The following is a typical scenario: user u sitting on machine M wishes to obtain a set S of images pertaining to, say, the internals of a computer. The user's query Q is sent to a high-level intelligent agent A_0, which then tasks lower-level agents $A_1, A_2, \ldots, A_n$ on the basis of the information they are familiar with. In this multiagent environment, it is the responsibility of some agents to become very familiar with a very specific data type such as JPEG images. There are, in turn, more general or abstract agents that are responsible for a generic image type (which perhaps includes JPEG and GIF images). Ultimately, it is the job of the high-level intelligent agent A_0 to present the user with the results of a specific web query which may include several images of differing image types. Rahimi and Carver in [5] expose a similar design.

5.4.3 Mobile Agents

The distinguishing factor between an agent and a mobile agent lies in the fact that mobile agents possess the capability to migrate from one peer to another autonomously. They are usually modeled as finite state automata. Typically, the migration process is performed as follows (with respect to the agent) [8]:

- *Suspension of execution*: The mobile agent's chain of execution is suspended at the onset of some predetermined event.

- *Identification of current state*: For later continuation, the agent's current state is saved.

- *Serialization*: The agent is internally converted to an array of bytes for transit.

- *Encoding*: The byte array is encoded (and perhaps compressed) to some common format used in the communication process.

- *Transfer to a remote host* (*subsequent to host authentication*): The encoded agent is transferred, via the network, to the host.

With respect to the agency, the process is only slightly different:

- *Authentication of the communication*: The host authenticates the communication (the previous host's request to transfer the agent).

- *Decoding*: The incoming data are decoded back to an array of bytes.

- *Deserialization*: The byte array is deserialized to the object form of the agent.

- *Instantiation*: The object is then instantiated into the mobile agent.

- *Restoration of previous state*: The agent's state is set to the previously saved state in order to properly resume execution on this host.

- *Continuation of execution*: The agent now resumes execution, its current state guiding its flow of control.

5.4.4 Web Services

A web service is a software system that is designed to provide a reliable mechanism for machines to interact over a network [9]. It provides universal interoperability between platforms and architectures, applications, and languages using commonly utilized standards. For example, it features communication over the standard Hypertext Transfer Protocol (HTTP), and the most common system implementation includes clients (applications) and servers (web services). Web services further support a remote procedure call (RPC), allowing an application method to be "called" remotely, and SOA—where the message is more important than the operation behind the message. They utilize XML messages that follow the Simple Object Access Protocol (SOAP) standard for exchanging structured information. With web services, it is possible to break an application into many parts and distribute them remotely; the application can then be executed on numerous remote systems that host its parts.

A web service publishes its service description in a machine-readable format known as the Web Service Description Language (WSDL). The discovery of available web services is achieved via a Universal Description Discovery and Integration (UDDI) registry, essentially a global "phone book" of web services. Discovery of mobile agent–enabled web services is then easily performed via a simple lookup of the registry.

Within the hyperdistribution environment, web services provide the capability to implement markets. They are key to the efficient hyperdistribution of contextual information because they provide a straightforward method to distribute producers and consumers and to get them together to exchange information.

5.4.5 Security Issues with Web Services

Utilizing autonomous intelligent agents to perform tasks without user intervention merits hefty security protocols. Generally speaking, security threats become increasingly significant in the presence of mobile code (e.g., mobile agents). Particular to mobile agents is their intrinsic vulnerability once situated on a host. Typically, the mobile agent must give access to its code, potentially its state, and at times its data. This poses a difficult problem for mobile agent designers. Perhaps, for example, the owner of a mobile agent does not wish to release its code to an agency (i.e., it is of a proprietary nature). Numerous methods have been proposed in answer to this problem, although none are comprehensive, and none propose a complete security strategy.

The issue of security threats regarding networked environments (e.g., LAN and WAN) and communications in general has been discussed extensively [10, 11]. In fact, many of the security protocols that are currently implemented are strictly based on the following set of assumptions [11]:

- *Identity assumption*: In this assumption, we can identify the person who wishes to perform some action. With this information, we can determine whether to allow the action to take place. Identities are typically verified via the use of user names and passwords (e.g., on Linux and Windows operating systems). Frequently, computer programs perform tasks on behalf of an individual.

- *Trojan horses are rare*: All programs are generally identifiable and are from trusted sources. Clearly, mobile agents as defined are in direct violation of this. This assumption may indeed be questioned of late due to the enormous amount of Trojan horses and worms lurking around today on the Internet.

- *Origin of attacks*: The prototypical hacker is used to model security. Security threats come from attackers running programs with malicious intent; therefore, security measures should authenticate the user and make sure that programs run by this user can only do those things that the user is allowed to do.

- *Programs stay put*: In general, programs rarely cross the administrative boundary. When they do, it is intentional. A program runs on one operating system (OS), and security is provided by the OS. Computer programs do not move from one system to another unless users intentionally transmit them.

Mobile code violates all these security assumptions to the point that even single-user systems require hefty security in the presence of mobile code. Characteristics specific to mobile agents further present security risks and thus motivate the continuous work done with respect to their security. The ongoing push toward MAS further complicates security issues, particularly with respect to concurrent distributed systems. In general, security issues with respect to mobile agents lie in three classes [12]:

- *Security of the agent*: An agent may be compromised by a malicious agency; furthermore, an agent may be operating in an unfriendly environment and thus be vulnerable. Typically, mobile agents can be adversely affected by malicious hosts and other malicious agents.

- *Security of the agency*: An agency may be compromised by malicious agents or some other entity altogether. It is typical to involve the underlying operating system and/or supporting software (e.g., firewalls) in the security of the host.

- *Security of the communication channel*: Agent–agency communication is performed via communication channels that are at risk of being compromised. Often, sensitive data are transferred using these channels.

It is thus necessary to consider several cases: first, the agency must be protected from malicious agents and other malicious entities that may target it. Second, agents must themselves be protected from several potential types of attackers. Malicious agencies may attempt to compromise agents in a number of ways (e.g., denial of service or eavesdropping on collected data), and other malicious agents may attempt to convert an otherwise trustworthy agent and force it to do its bidding. Third, the data that the agent carries in its *briefcase* must also be protected. Quite often, the data are of a proprietary nature (particularly in secure networks) and must not be compromised.

There are many issues to ponder when considering the security of mobile agents and web services. More than likely, a web service will know nothing about the agents it authorizes, so it must be protected from potentially malicious agents that try to steal data or have access to resources they should not have access to. In the typical case, it will provide agent identification, authentication, verification, and execution restriction within the "sandboxed" environment. A commonly utilized technique involves tracing an agent's execution while on the host [13, 14]. The trace can later be analyzed for anomalies.

The reverse is also true in that a malicious web service can pose as a legitimate one, thereby potentially harming an agent by attempting such things as reverse engineering, stealing data, or preventing the agent from arriving at its intended destination(s). Far worse may be to redirect and reprogram an agent to perform tasks that it was not originally intended to do. This can be especially problematic for agents used in e-commerce; for example, consider an agent whose job is to purchase a service on your behalf. It may be reprogrammed to purchase the most expensive service or perhaps the wrong service altogether. This remains an important issue that precludes the mainstream use of mobile agents in e-commerce.

There are numerous techniques one can implement when attempting to provide a secure environment for mobile agents. For example, it may be possible to restrict access to a proven trusted environment. Unfortunately, this is particularly difficult to accomplish and is further encumbered by a lack of formal tools to ensure provability in software. Secure hardware may be used, but it is expensive and therefore not particularly practical. Encryption of the agent while in transit provides adequate security during the migration process, but offers no protection while the agent is on the host. There have also been many techniques that attempt to provide partial agent security; for example, partial result authentication codes [12] encrypt execution results on a host with a throwaway key prior to migrating to another host. Malicious agencies cannot then forge results from previous hosts. Encrypted functions and obfuscated code [13] attempt to hide proprietary algorithms. Environmental key generation [12] assures the agent that it is executing on a trusted host by generating a key that is strictly dependent upon its executing environment; a malicious host may not know the specifics of that environment. The master–slave approach exploits the idea that many slave agents do nominal work and are not aware of the scope of the entire task. If they are compromised, the leak of information is contained and minimized.

These techniques ultimately depend on the agent designers' requirements—what is more important to them. In some cases, the data an agent gathers or the information that is inferred from processing the data must be protected from prying eyes. In other cases, an agent's itinerary (where it has been and is subsequently going) must not be disclosed; or perhaps its algorithms are proprietary and state of the art. In any case, numerous techniques exist that attempt to address these and other security requirements [12].

Several other formal methodologies also address web service security. WS-Security [15, 16, 17] provides XML encryption and XML signature in SOAP. Extensions to this such as WS-Trust and WS-SecureConversation [15] further extend it by addressing the security of entire web service sessions as opposed to individual messages. WS-Security was originally designed to augment Hypertext Transfer Protocol Secure (HTTPS). WS-Reliability [18] is another standard for reliable interweb service messaging, and WS-Transaction standardizes the coordination of web service transactions.

5.4.6 The Use of Web Services as Mobile Agent Hosts

Web services are inherently static and adhere to a typical synchronous paradigm. On the other hand, MASs follow an asynchronous methodology in that there is no need to maintain connectivity of any kind. Our architecture consists of intelligent mobile agents and web services that replace traditional agencies. The web service becomes the agency and provides the agent's interface to services. XML/SOAP provides the transport mechanism. It is possible to implement web service security standards such as WS-Security to achieve secure migration.

Migration is achieved by "passing" a mobile agent in an encrypted serialized form as a web service method parameter. In this case, it is ultimately packaged in the XML/SOAP message and encrypted. Resumption of mobile agent execution on a new host is achieved via decryption and deserialization of the agent with subsequent instantiation and state appraisal. So migration essentially simplifies to a form of RPC whereby the mobile agent is a "method" parameter.

In the context of hyperdistribution, web services would function in this way as markets, providing communicability and support for producers and consumers.

5.4.7 Security Issues with the Use of Web Services as Mobile Agent Hosts

This architecture supports the use of existing mobile agent security techniques in addition to typical encryption methods provided by, for example, WS-Security. Proprietary encryption algorithms can be implemented by encrypting each serialized agent prior to migration and subsequently decrypting it. Other techniques can be implemented by providing the necessary methods on the web service.

Security in transit can be accomplished by using a variety of encryption algorithms. Each agency generates a unique public and private key pair. Prior to migration, the agent generates a random symmetric key and initialization vector (IV) which is used to encrypt the agent. The agent then uses the agency's public key to encrypt this key and IV. It then migrates (along with its encrypted key) to the agency via our method. The agent's key is then decrypted by the agency using its own private key, and the agent is subsequently decrypted using the decrypted key and IV.

The web service must then authenticate the decrypted agent. This is accomplished by simple type verification. The process of deserialization and instantiation cannot successfully occur unless the agency maintains a valid genetic makeup of the agents it will accept and authenticate.

Identification of the agent is performed via a biometric identifier. This identifier is verified by the sandbox. In the basic case, it can simply be a hashed key which is simply matched to some stored version, thus implying the use of a database of supported individual agents. Authorization of resource use and data access is achieved by further database lookup to determine approved operations.

5.4.8 Web Services as Static Agents

There are cases in which static agents prove to be useful. For example, Rahimi and Carver propose a hierarchical multiagent architecture that employs the use of information agents, the goal of which is that each agent provides expertise on specific topics [5]. Some agents, called *wrappers*, simply provide knowledge of a very specific type of information source (e.g., relational databases, images, text, and audio sources). These wrappers aid in the identification of data types and services, and are considered static (i.e., they do not migrate).

Web services as agencies can be retooled to function as static agents. Simply exposing a method that performs the task of a static agent is adequate. Other agents (perhaps mobile) can then query the static agent by "calling" this method and consuming the web service. The result expresses a reply by the static agent.

In the context of hyperdistribution, web services would function in this way as nonnomadic producers (i.e., they do not migrate but instead remain fixed on a particular server). They may be experts in contextual information belonging to a subset of the global information scope. They continuously mine information from a local data repository and update their context in order to provide it to migrating consumers when requested.

5.4.9 Hyperdistribution Methods

Hyperdistribution is made possible through the use of independent, autonomous mobile agents that serve as producers and consumers. They migrate from market to market, peer to peer, and data source to data source. Producers collect contextual information. Consumers locate producers with the contextual information they need in order to perform some action. Discovery of producers, consumers, and services is made possible through the markets. Consumers can advertise their need, and producers can provide the contextual information. Services such as access to data and computational resources are also advertised and provided via the markets.

Consider a simple hyperdistribution example: a consumer has some information it needs context for in order to determine if an action should be performed. It knows of a single market located at some IP address—to which it migrates to see if it can find the contextual information it requires. Noticing no producers around the market, it decides to search the local database first. Suppose it finds nothing useful. The consumer then decides to post its needs on the message board for some producer to notice the next time it arrives at this market. It also notices the location of several new markets that have popped up in the network. Being a persistent consumer, it decides to visit these new markets; however, in order to be contacted as soon as a producer comes along that can provide the appropriate contextual information it needs, it provides its itinerary for the near future. From this point, many scenarios can ensue. For example, a producer may come along, read the message board, and migrate to the locations provided in the consumer's itinerary. The producer may encounter the consumer, and the information may be communicated. Or perhaps it may simply leave the contextual information for the consumer via the message board. At a later time, the consumer may revisit the market and obtain the context there. It is important to note that markets are not the only entities that can provide producer–consumer communicability; at any point on the network, producers and consumers may "meet" and communicate.

5.5 MODELING TOOLS

In the previous sections, we have examined an abstract agent model for the hyperdistribution of contextual information. It is useful for hyperdistribution and contextual processing to have formal methods that can be used to model the system. These formal methods provide proof of system integrity in that they can assure us that it will function as intended. Furthermore, they allow us to quantify system scalability and performance prior to embarking on the work that is involved in realizing it. In this section, we provide a brief overview of a selection of modeling tools that aid in the design of agent-based systems. It is important to note that these tools provide a method of modeling a large number of systems—hyperdistribution being only one of them. As such, they are fairly abstract, but possess powerful constructs and primitives.

5.5.1 π-Calculus

π-Calculus is one of the most flexible calculi and one that allows us to model communicating processes such as producers and consumers in the hyperdistribution of contextual information. It is a process algebra which has its roots in calculus of communicating systems (CCS) [19], which was originally developed in the late 1970s. In fact, it is an extension of CCS that supports mobility. In general, process algebras define variables which denote arbitrary channels and processes (operators); operators describe relations between processes and channels [20]. π-Calculus models the changing connectivity of interactive systems, focusing on the interaction between processes. It provides primitives for describing and analyzing distributed systems which focus on process migration between peers, process interaction via dynamic channels, and private channel communication. Example applications include languages that support distributed programming with process mobility and description and analysis of authentication protocols. For example, in hyperdistribution the calculus would allow us to model the communication of contextual information from producer to consumer.

What follows is an informal introduction to the π-calculus. For a more thorough review, the reader is referred to work presented by Milner et al. [21, 22] and Parrow [19].

5.5.1.1 Overview

According to Wing [23], a *process* is an abstraction of an independent thread of control. A *channel* is an abstraction of the communication link between two processes. It is simply a way for two processes to communicate. For example, in hyperdistribution it may simply be a socket shared by the producer and consumer. Processes interact with each other by sending and receiving messages over channels. What essentially separates π-calculus from other modeling methods is the ability to pass channels as data along other channels; this allows the expression of process mobility, which allows expression of changes in process structure. Therefore, π-calculus is a natural choice for describing concurrent processes that communicate through message passing. However, it is not good at describing abstract data types or states with rich or complex data structures.

5.5.1.2 Preliminary Definitions

π-Calculus consists of a set of prefixes and process expressions [19, 20, 23]. The basic building block is the set of infinite names N which vary over a, b, ... z. These names function as communication channels, variables, values, and process identifiers. The prefixes are as follows:

$\bar{a}\langle x \rangle$	Output	The message x is sent along the channel a.
$a(x)$	Input	The channel a can receive input and bind it to x.
τ	Silent	Nothing observable happens.

In general, α denotes an arbitrary action prefix (*input*, *output*, or *silent*). Agents (or processes), varying over P, Q, ..., are defined in Table 5.1, where ::= means *is defined to be*. Agents can be of the following form:

- *0*, the nil process (empty process), which does nothing. In hyperdistribution, the nil process may model a consumer that has yet to be discovered.

- $\bar{a}\langle x \rangle.P$, an *output prefix*; the process sends a message x over the channel a and then proceeds as P. In hyperdistribution, a producer P may send contextual information i to the consumer on some communication channel L as follows: $\bar{L}\langle i \rangle.P$.

- $a(x).P$, an *input prefix*; the process waits on channel a to receive a message that is then bound to x and then proceeds as P. In

TABLE 5.1 π-Calculus Syntax

Prefixes	α	$::=$	$\bar{a}\langle x \rangle$	Output
			$a(x)$	Input
			τ	Silent
Processes	P	$::=$	0	Nil
			$\alpha.P$	Prefix
			$P+P$	Sum
			$P \mid P$	Parallel
			if $x = y$, then P	Match
			if $x \neq y$, then P	Mismatch
			$(vx)P$	Restriction
			$A(y_1, y_2, \ldots, y_n)$	Identifier
Definitions			$A(x_1, x_2, \ldots, x_n) \underset{=}{\mathrm{def}} P$	$(\text{where } i \neq j \Rightarrow x_i \neq x_j)$

hyperdistribution, a consumer C may receive contextual information i from a producer on some communication channel L as follows:
$L(i).C$.

- $\tau.P$, the *silent prefix*; nothing observable happens (i.e., the process
  can proceed as P by doing nothing).

- $P+Q$, a *sum*; nondeterminism (i.e., the process can represent either
  P or Q). In hyperdistribution, this could model two producers that
  possess the information necessary for a consumer; any one of the
  two can provide this information in a nondeterministic manner.

- $P|Q$, a *parallel composition*; processes P and Q run concurrently. In
  hyperdistribution, parallelism is plentiful. A producer P and con-
  sumer C may run concurrently as $P|C$.

- If $x = y$, then P, a *match*; if x and y are the same name, the process
  will behave as P. In hyperdistribution, a consumer's rule may apply
  if a match occurs on two pieces of information (e.g., x and some context y); the action associated with that rule may then be performed.

- If $x \neq y$, then P, a *mismatch*; if x and y are *not* the same name, the
  process will behave as P. This is the inverse of a match.

- $(\nu x)P$, a *restriction*; the variable x is bound to P (i.e., it is local and cannot be immediately used for communication between P and its environment). For example, this may represent contextual information possessed by some producer P that is unknown to any other producers or consumers within the same environment.

- $A(y_1, y_2, \ldots, y_n)$, an identifier; n is the arity (number of parameters) of the function A, and every identifier has a definition $A(x_1, x_2, \ldots, x_n) \underline{\text{def}} P$, where x_i are pairwise distinct. Every argument to A is unique.

Match and mismatch constructs are the only valid tests within the calculus; they can only compare names for equality. They can be combined (including by using sum); for example,

$$\text{If } x = y, \text{ then } P + \text{If } x \neq y, \text{ then } Q$$

behaves as P if $x = y$ or Q if $x \neq y$ is true. Note that this example can be abbreviated (thus simplified) by the following:

$$\text{If } x = y, \text{ then } P, \text{ else } Q.$$

Furthermore, the following shortcuts are often employed for the *match* and *mismatch* operators:

$$\text{If } x = y, \text{ then } P \equiv [x = y]P.$$

$$\text{If } x \neq y, \text{ then } P \equiv [x \neq y]P.$$

The input prefix $a(x).P$ binds x in P; however, the output prefix $\overline{a}\langle x \rangle.P$ does not. Input and output prefixes have a subject a and an object x, where the object is *free* in the output prefix and *bound* in the input prefix. *Bound names* in P are defined as bn(P), and *free names* in P are defined as fn(P).

Substitutions are maps for names. For example, the substitution that maps y to x is written as x/y. In general, $x_1, x_2, \ldots, x_n/y_1, y_2, \ldots, y_n$ for pairwise distinct y_i maps each y_i to x_i. σ is used to range over all substitutions, and sometimes $\tilde{x}$ can be used to represent a sequence of names of unimportant length.

5.5.1.3 The Polyadic π-Calculus

The polyadic π-calculus allows the communication of more than one name (e.g., tuples of names, sorts, functions, and data structures) during a single action [19, 20]. For example:

$$\bar{a}\langle b_1, b_2, \ldots, b_n \rangle.P | a(x_1, x_2, \ldots, x_n).Q$$

which can be encoded in the monadic π-calculus as follows:

$$(vw)\bar{a}\langle w \rangle.\bar{w}\langle b_1 \rangle \ldots \bar{w}\langle b_n \rangle.P | a(w).w(x_1).w(x_2) \ldots w(x_n).Q$$

In general:

$$a(\tilde{x}).P | \bar{a}\langle \tilde{y} \rangle.Q \xrightarrow{\tau} P\{\tilde{y}/\tilde{x}\} | Q$$

The mobility represented by the polyadic π-calculus implies a reference passing, wherein processes do not actually move; instead, the communication configuration changes [24]. In the context of hyperdistribution, this would model a producer sending multiple pieces of contextual information to a consumer.

5.5.2 Ambient Calculus

Ambient calculus introduces the concept of an *ambient* in order to model the movement of one location of activity inside another. An ambient is a fundamental primitive in the calculus and is simply a bounded place where computations can occur—it is the key to representing mobility in the calculus. In the context of hyperdistribution, ambients might allow us to model markets in which many producers and consumers communicate contextual information amongst each other. For example, a model containing two ambients may represent two unique markets. Then, using the calculus, we may model migration between the two markets and communication from producer to consumer within the markets. Ambient calculus groups processes into multiple, disjoint, distributed ambients (as opposed to π-calculus); enables interaction by shared position; and implements access control by using capabilities derived from ambient names [25]. Security in the ambient calculus is implied by the view that crossing boundaries is guarded. The mobility represented in the calculus is general in that it allows mobility of processes, channel names, and entire environments [24]; the crossing of boundaries implies mobility. Boundaries surrounding ambients permit the definition of separate locations, thus inferring some sort of abstract knowledge of the distance between locations. Interagent communication occurs within a common boundary.

What follows is a cursory introduction to the ambient calculus. For a more thorough review, the reader is referred to work done by Cardelli and Gordon in [25–28].

5.5.2.1 Ambients

In general, communication within an ambient is local to that ambient, is anonymous, and is asynchronous. Ambients have names which strictly control access (e.g., *entry*, *exit*, and *communication*). They can be nested within other ambients to produce computational environments within computational environments. Moreover, ambients can be *moved* entirely through channels defined in the calculus. Boundaries surrounding ambients rigorously control what is inside and outside the ambient. Examples of ambients include a web page (bounded by a file), a laptop (bounded by its case) [26], and, in the case of hyperdistribution, a consumer (bounded by a market). Each ambient has, within it, agents which perform computations.

5.5.2.2 Mobility and Communication

The primitives of the calculus are given in Table 5.2, where ≡ means *is defined to be*.

The restriction operator $(vn)P$ creates a new unique name n within a scope P. This name can then be used to name ambients and to operate on ambients by name.

The process 0 does nothing. An ambient $n[P]$ possesses name n with process P running inside of it. It is implied that P is *actively* running and that P can be made of more than one concurrent process. In hyperdistribution, this might model a producer P in a market n. Since ambients can be nested within each other, it is entirely possible to have, for example, $n[P|m[Q]]$. The process $M.P$ executes an action regulated by the capability M, then continues as the process P. Note that P is not running until the action is executed. In hyperdistribution, we might model a producer P and consumer C at the same market M as follows: $M[P|C]$.

The entry capability *in m.P* instructs the ambient surrounding *in m.P* to enter a sibling ambient named m or pictorially as shown in Figure 5.3.

TABLE 5.2 The Primitives of the Ambient Calculus

Processes

P,Q	$\equiv$	$(\nu n)P$	Restriction
		0	Inactivity
		$P\vert Q$	Composition
		$!P$	Replication
		$M[P]$	Ambient
		$M.P$	Action
		$(x).P$	Input action
		$\langle M \rangle$	Asynchronous output action

Capabilities

M	$\equiv$	x	Variable
		n	Name
		in M	Entry (can enter into M)
		out M	Exit (can exit out of M)
		open M	Open (can open M)
		ε	Null
		$M.M'$	Path

Conversely, the exit capability *out m.P* instructs the ambient surrounding *out m.P* to exit its parent ambient *m*. The open capability *open m.P* removes the boundary surrounding ambient *m*. In hyperdistribution, this may model groups of communicating producers and consumers.

Ambient I/O is possible via input and asynchronous output actions [26]. An output action releases a capability into the surrounding ambient.

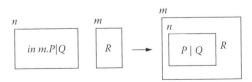

FIGURE 5.3 The ambient resulting from the process $n[in\ m.P\vert Q]\vert m[R]$.

Conversely, an input action requires a capability from the surrounding ambient which it binds to a variable. For example:

$$(x).P|\langle M \rangle \rightarrow P\{x \leftarrow M\}$$

where $P\{x \leftarrow M\}$ implies the substitution of the capability M for each free occurrence of the variable x in the process P. In hyperdistribution, this might model the posting of messages on the message board at a market.

5.5.2.2.1 A Mobile Agent Example   In the following example, a mobile agent wishes to leave its home and later return. The home host must, however, authenticate the mobile agent before it can resume execution:

$Home[(vn)(open\ n\ |\ Agent[out\ home.in\ home.n[out\ Agent.open\ Agent.P]])]$

In this expression, *Home* may represent an agent's home agency and (vn) models a secret n known only to the agent. The agent initially resides within its home agency. Its goal is to migrate from there, distribute the secret over the network, and return. The expression *out.home* models the agent's migration out of the home agency; *in.home* models the agent's return to its home agency. What occurs prior to returning home is not modeled in this case. The expressions *out Agent* and *open n* allow the secret to be shared to all within the home agency; finally, *open Agent.P* merely allows the agent to then behave as some process P. Essentially, a shared secret n within the safe location *Home* is distributed over the network (carried by *Agent*); the authentication is based on this shared secret [26]. The following is the behavior of the computation:

$Home[(vn)(open\ n\ |\ Agent[out\ home.in\ home.n[out\ Agent.open\ Agent.P]])]$

$\equiv (vn)Home[open\ n|Agent[out\ home.in\ home.n[out\ Agent.open\ Agent.P]]]$

$\rightarrow (vn)(Home[open\ n\|Agent[in\ home.n[out\ Agent.open\ Agent.P]])$

$\rightarrow (vn)Home[open\ n|Agent[n[out\ Agent.open\ Agent.P]]]$

$\rightarrow (vn)Home[open\ n|n[open\ Agent.P]|Agent[]]$

$\rightarrow (vn)Home[0|open\ Agent.P|Agent[]]$

$\rightarrow (vn)Home[0|P|0]$

$\equiv Home[P]$

We can adapt this example to hyperdistribution: the behavior exhibited by the expressions similarly models a consumer (*Agent*) migrating to a market (*Home*) and placing a request for contextual information (*n*) on a message board located within the market. The ambient *n* is simply modeling the consumer's message. The action represented by *n* exiting the ambient *Agent* and subsequently *open*ing models the posting of this message to the message board. In the end, the consumer (modeled as the process *P* in the above expressions) remains at the market.

5.5.3 Petri Nets

Petri nets offer a pictorial view of the system [29]. They are generally used to model systems that exhibit inherent parallel, asynchronous, and concurrent characteristics. In the context of hyperdistribution, they could model a system and grow dynamically as the system changes; furthermore, they could be used to model specificities involving routing, process connection, and control flow. In this section, an informal introduction to Petri nets is presented. Various Petri net extensions are further discussed. For a formal and thorough review of Petri nets, the reader is referred to [30].

5.5.3.1 Overview

Petri nets are pictorially represented as a bipartite graph composed of two main components. Figure 5.4 illustrates a typical Petri net. Nodes are

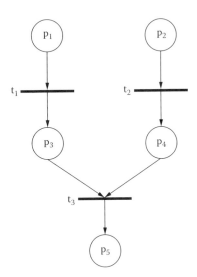

FIGURE 5.4   Graph representation of a Petri net.

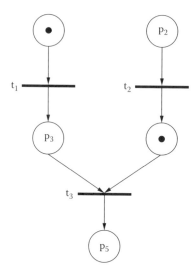

FIGURE 5.5   A marked Petri net.

composed of *places* (the circles in the figure) and *transitions* (the bars in the figure) which are connected by directed *arcs*. These arcs may denote direction from a place to a transition or vice versa. An arc that goes *out* of a node i and *into* a node j (the configuration of these nodes—i.e., which is a place and which is a transition—is unimportant at the moment) denotes that i is an input to j and that j is an output of i [31]. Transitions have an input place and an output place.

The dynamic properties of a Petri net can be modeled with the use of movable tokens that represent a marking (state), thus yielding a marked Petri net as shown in Figure 5.5. A marked Petri net is said to have an initial marking. A place may have any number of tokens. Several rules must be followed with respect to moving the tokens around [31]:

- Tokens are moved by the *firing* of transitions.

- A transition must be *enabled* in order to fire.

- A transition is enabled if all of its input places contain a token.

- The firing process removes a token from each of a transition's input places and puts a token in each of a transition's output places. This occurs atomically.

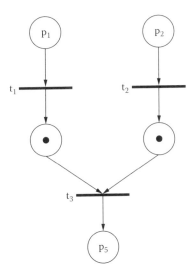

FIGURE 5.6   The Petri net resulting from the firing of transition t_1.

Multiple transitions may be enabled simultaneously, any one of which can fire. Note that it is not mandatory that an enabled transition fire. Moreover, firing is nondeterministic, thus Petri nets are well suited for modeling the concurrent behavior of distributed systems. Producers and consumers within the hyperdistribution environment are nondeterministic; they do not necessarily behave predictably. In the original definition, only one token could be removed from and added to a place during the firing of a transition. Weighted Petri nets have since been defined which generalize the original and allow multiple tokens to be removed from and added to a place. Typically, arcs are labeled with the appropriate weight; unlabeled arcs imply a weight of 1.

In Figure 5.5, only transition t_1 is enabled because it has a token in all of its input places (place p_1). Transition t_3 is *not* enabled because it does not have a token in one of its input places (place p_3). When transition t_1 fires, the marking changes (the token in p_1 is removed, and a token is placed in p_3) to the Petri net shown in Figure 5.6. Note that transition t_3 is now enabled because all of its input places have a token (places p_3, p_4). Once transition t_3 fires, the Petri net transitions to that of Figure 5.7.

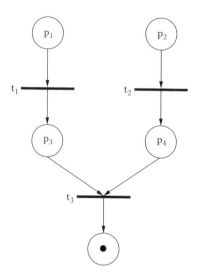

FIGURE 5.7   The Petri net resulting from the firing of transition t_3.

5.5.3.2 Formal Definition

A Petri net is defined as a 5-tuple (P, T, A, M_0, W), where

$P = (p_1, p_2, \ldots, p_m)$ is a set of places;

$T = (t_1, t_2, \ldots, t_n)$ is a set of transitions;

$A \subseteq (P \times T) \cup (T \times P)$ is a set of arcs wherein no arc may connect two places or two transitions;

$M_0 : P \rightarrow N$ is an initial marking; and

$W : A \rightarrow N^+$ is a set of arc weights, which assigns to each arc $a \in A$ some $n \in N^+$ specifying the number of tokens a transition consumes from a place or puts in a place when it fires.

The marking of a Petri net can be stated as a vector M, where the first value in the vector represents the number of tokens in place p_1, the second value represents the number of tokens in place p_2, and so on. For example, the marking of the Petri net in Figure 5.5 can be stated as $M = (1,0,0,1,0)$. The *state space* of a Petri net is the set of all possible markings. The *reachability tree* of a Petri net is the set of all states that the net can enter into by any possible firing of a transition. Often, Petri nets suffer from what is referred to as the *state space explosion problem*; there are too many markings that

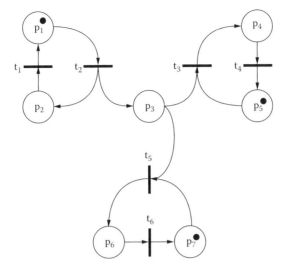

FIGURE 5.8   The Petri net modeling a simple producer–consumer scenario.

the net can take on, thus rendering it difficult to analyze. Several techniques (e.g., in [32]) have been proposed related to this problem.

Petri nets can be used to model quite a large number of things from
the elementary to the most complicated. For example, a traditional producer–consumer problem is modeled in Figure 5.8, and the equation
$x = (a+b)/(a-b)$ is modeled in Figure 5.9. Clearly, these are very simple
examples to model.

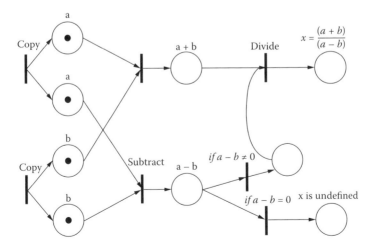

FIGURE 5.9   The Petri net modeling the equation $x = (a+b)/(a-b)$.

5.5.3.3 Extensions to the Petri Net

Removing the single token limitation was one of the first extensions to the simple Petri net. *Generalized Petri nets* allow the removal of more than one token from an input place and the addition of more than one token to an output place upon the firing of a transition.

Mobile Petri nets [33] are a generalized form of Petri nets and are composed of mobile nets and dynamic nets. *Mobile nets* support the passing of a reference to a process along a communication channel. Communication channels are modeled as places in the net, and mobility is implemented by the moving of colored tokens (tokens with values). *Dynamic nets* provide the capability to instantiate a new net during the firing of a transition for true dynamic support. Mobility, in this case, can be viewed as the process of creating a copy at the new location and terminating execution at the current location [24].

Communicative and cooperative nets [24] provide mechanisms that allow Petri nets to communicate with each other. Effectively, *communicative nets* provide the capability to model communication via message passing, and *cooperative nets* model communication via the client–server paradigm. In communicative nets, a subset of places is reserved for data that are scheduled to be sent to another net. Transitions are relatively more involved, personifying data function calls similar to a producer–consumer, message-sending, and object creation. Cooperative nets substitute message-sending transitions for service request and retrieve transitions.

Concurrent object-oriented Petri nets (CO-OPN/2) [24, 33] allow the modeler to deal with interacting object-oriented systems, including many traditional facets of the object-oriented model (e.g., inheritance, abstraction, and polymorphism). Many object (algebraic) nets interact with each other. ■

5.6 ADVANCED TOPICS

In this section, our aim is to discuss topics that might be addressed when considering the hyperdistribution model presented in this chapter. These include methods that assist in the modeling of hyperdistribution of contextual information. The eventual goal of such a discussion would be the creation of a model where information flows to where it is needed and can be edited and disseminated by interested parties based on trust relationships.

5.6.1 API-S Calculus

The API-S calculus (API-S) is an extension of the π-calculus and API calculus [34] that is designed to address some of the limitations of these calculi for the purpose of modeling intelligent agents. Since our hyperdistribution model utilizes agents as producers and consumers, it behooves us to discuss modeling tools that are specifically designed to model such entities and their unique characteristics. Specifically, API-S introduces primitives that provide the necessary mechanisms to model the security characteristics of agent-based systems. It also addresses the natural grouping behavior of cooperating agents, something that is frequently encountered in the hyperdistribution of contextual information. API-S forms a foundation on which additional constructs can be built, thus offering the necessary tools to extend the breadth of the types of systems the calculus can model. In hyperdistribution, API-S could be used to model producers and consumers (their migration and communication), including the security implemented by both (e.g., while migrating, for information hiding and encryption).

API-S further expands upon several concepts defined in the API calculus in order to facilitate the modeling of numerous security techniques and protocols. The security of the agent and of the agency are both considered. API-S provides mechanisms to completely secure actions involving the knowledge of agents within a multiagent system. As in API calculus, agents possess the ability to alter their knowledge (via knowledge units) by adding to or dropping facts from their fact list and to alter the knowledge base by adding or removing specific rules to their rule base. An agent may possess one or more knowledge units, and support for sending such units to or receiving them from other agents is inherently provided in a secure fashion. This provides the capability to model the communication of contextual information from producer to consumer. Moreover, API-S introduces new primitives that allow modeling a wide variety of cryptographic protocols.

API-S can model a group of agents working cooperatively on a similar task or even a group of agents residing on the same agency involved in secure communication. This proves useful within the context of hyperdistribution, particularly with respect to markets and groups of cooperating producers and consumers.

5.6.1.1 Syntax

The following constitutes the basic definition for the elements of API-S.

5.6.1.1.1 Term

R,T	$::=$	$x,y,z,...$	Name
	$\|$	$a,b,c,...$	Fact or rule
	$\|$	$f(x,y,z,...)$	Function
	$\|$	R^{Ω},T^{Ω}	Ω-Term

A name may be a channel of communication or simply a variable name. A fact or a rule can be added to a knowledge base, sent to a different agent, or received from a different agent. A function may have l parameters; f ranges over the functions of Φ, and one matches the arity of f. In hyperdistribution, a name is simply a variable representing any number of communicable entities. A fact or rule allows us to model a knowledge base and thus the intelligence of an agent. A function can represent any number of behaviors we may wish to hide details of; for example, data gathering and filtering. An Ω-term (crypto-term) allows us to incorporate security aspects.

Variables u, v, and w are used to range over names, facts, and rules. The abbreviations $\vec{u}$ and $\vec{T}$ are used in place of tuples $u_1,u_2,...,u_l$ and $T_1,T_2,...,T_l$, respectively. Terms are partitioned into a collection of *subject sorts*, each of which contains an infinite number of terms. We write $R{:}s$ to mean that term R belongs to subject sort s. As in the API calculus, this notation is extended to tuples component-wise.

Object sorts, ranged over by S, are simply sequences over subject sorts, for example $(s_1,s_2,...,s_l)$ or just (s). A sorting is a function Sf mapping each subject sort to an object sort. We write $S \rightarrow (\vec{s}) \in Sf$ if Sf assigns the object sort $(\vec{s})$ to s. In this case, we say that $(\vec{s})$ appears in Sf. By assigning the object sort (s_1,s_2) to the object sort s, the object part of any term in s is forced to be a pair whose first component is a term of s_1 and whose second component is a term of s_2 [35].

An Ω-term compartmentalizes cryptographic primitives for use in modeling an array of cryptographic protocols and techniques.

5.6.1.1.2 Process

P	::=	0	No action
	|	$\alpha.P$	Action prefix
	|	$P_1 + P_2$	Summation prefix
	|	$[T = R]P_1 : P_2$	Conditional prefix
	|	vxP	Name restriction
	|	$(K_i)P$	Knowledge unit restriction
	|	$!P$	Replication
	|	$D\langle \vec{L} \rangle$	Constant
	|	$M\langle\langle P \rangle\rangle N$	Listener
	|	P^Ω	Ω-Process

Processes are denoted by $P_1, P_2, \ldots, P_n$ and $Q_1, Q_2, \ldots, Q_n$. The no-action process is denoted by 0 and performs some internal computation. This is useful if we wish to hide the details of some action that is not necessary to model. The prefix α is called the *action prefix*. The expression $\alpha.P$ performs the action α (e.g., a communication of contextual information from producer to consumer) and then behaves as P. The summation process $P_1 + P_2$ behaves as either P_1 or P_2 and is nondeterministic. Terms are identified by T and R. The process $[T = R]P_1 : P_2$ is then a conditional process wherein the equality of T and R is not a strict syntactic identity. When $P_2 = 0$, we abbreviate the conditional process to $[T = R]P_1$. The expression vxP makes a new private name x local to P, then behaves as P. This provides scope restriction. The knowledge unit restriction $(K_i)P$ makes a new private knowledge unit K_i local to P. The replication process $!P$ implies $(P|P|\ldots)$; for example, $P_1|P_2$ consists of two agents, P_1 and P_2 acting in parallel and perhaps independently. Process synchronization (i.e., of an output action of P_1 at some output port $\bar{x}$ with an input action of P_2 at x) is possible, resulting in a silent τ action.

The constant D, where $\vec{L}$ indicates a tuple of processes or terms, is a function whose parameters can be processes or other functions. For example, consider $\bar{x}L$ to be an output prefix that sends a term or process L on channel x, and consider $x(L)$ to be an input prefix that receives a term or process L from channel x. Now consider the expression $\bar{x}P.Q|x(X).X$. Once the interaction between the two processes has occurred, the resulting process is then $Q|P$. This may be interpreted as a process $x(X).X$ waiting for X to be sent along channel x, thus defining its subsequent behavior. In this case, the process P is sent along channel x in the expression $\bar{x}P$; the process $x(X).X$ then simply behaves as P once the interaction has taken place.

In $M\langle\langle P\rangle\rangle N$, process P is associated with (*listening* to) milieus M and N. A *milieu* (defined later) is simply a bounded place where computations can occur. We may, for example, model a market using a milieu. Within the market, many producers and consumers may communicate; we can model this by using milieus. The milieu listener allows a process to be part of more than one communication group. Process P may receive messages broadcast by processes within M and N and may broadcast messages to any processes within M and N.

Similar to the Ω-term, we introduce the Ω-process (crypto-process) in order to allow us to formally reason about cryptographic processes. Ω-Processes provide support for a wide variety of cryptographic protocols, thus permitting secure analytical constructs in the calculus. This provides the tools necessary to model secure communication between producers and consumers in the hyperdistribution environment.

5.6.1.1.3 Ω-Term An Ω-term U includes primitives for cryptographic support.

R^Ω, T^Ω	::=	0	Zero
	|	(M,N)	Pair
	|	$suc(M)$	Successor
	|	$H(M)$	Hashing
	|	$\{M\}_K$	Shared key encryption
	|	$\{[M]\}_K$	Public key encryption
	|	$[\{M\}]_K$	Private key signature
	|	K^+	Public key part
	|	K^-	Private key part

Constructs for pairing and numbers, (M,N), 0, and $suc(M)$, are added for convenience in modeling numerous cryptographic protocols. The successor term $suc(M)$ implies $M+1$. For example, $suc(0)=1, suc(suc(0))=2$.

$H(M)$ represents the hashing of some message M (which can be a key, for example). Furthermore, we assume that H is a one-way function (i.e., it cannot be inverted) and that it is free of collisions. We write $\{M\}_K$ to represent the encryption of a message M with a key K using a shared key cryptographic system (e.g., DES [36]). Public key encryption (e.g., RSA [37]) is modeled as $\{[M]\}_K$, which illustrates some message M encrypted under some key K. Often, we wish to encrypt using some public key and typically write $\{[M]\}_{K^+}$ to denote this behavior. Accordingly, if K represents a key pair, then K^+ corresponds to its public part and K^- corresponds to its private part. We write $[\{M\}]_K$ to denote a message M that is digitally signed with key K. In hyperdistribution, this allows us to model the dissemination of encrypted contextual information, for example.

5.6.1.1.4 Ω-Process   To support the formal modeling of conditional processes regarding authoritative cryptographic protocols, the Ω-process is introduced. The digital signature of terms, processes, and agents is capably modeled in API-S, in addition to common cryptographic schemes.

$P^{\Omega} ::=$	$let\ (x, y) = M\ in\ P$	Pair splitting
$\mid$	$case\ M\ of\ 0 : P, suc(x) : Q$	Integer case
$\mid$	$case\ L\ of\ \{x\}_N\ in\ P$	Shared key decryption
$\mid$	$case\ L\ of\ \{[x]\}_N\ in\ P$	Public key decryption
$\mid$	$case\ L\ of\ [\{x\}]_N\ in\ P$	Signature check

In $let\ (x, y) = M\ in\ P$, the variables x and y are bound in P. It behaves as $P\{NL/xy\}$ if M is the pair (N, L). In $case\ M\ of\ 0 : P, suc(x) : Q$, the variable x is bound in Q. It behaves as P if M is 0 or as $Q\{N/x\}$ if M is $suc(n)$. The variable x is bound in P in all three of the following processes:

- $case\ L\ of\ \{x\}_N\ in\ P$ attempts to decrypt L with the key N. If L is a cipher text of the form $\{M\}_N$, then the process behaves as $P\{M/x\}$. This could, for example, model a consumer decrypting contextual information it has received with a shared key.

- $case\ L\ of\ \{[x]\}_N\ in\ P$ attempts to decrypt L with the key N. If L is a cipher text of the form $\{[M]\}_N$, then the process behaves as $P\{M/x\}$. If N is a private key K^-, then x is bound to M such that $\{[M]\}_{K^+}$ is L, if such an M exists. In hyperdistribution, this could model a consumer decrypting contextual information provided by a producer using public key decryption.

- *case L of* $[\{x\}]_N$ *in P* attempts to decrypt L with the key N. If L is a cipher text of the form $[\{M\}]_N$, then the process behaves as $P\{M/x\}$. If N is a public key K^+, then x is bound to M such that $[\{M\}]_{K^-}$ is L, if such an M exists. In hyperdistribution, this would provide support for modeling a signature check on some received contextual information.

We should note, quite importantly, that several assumptions about cryptography are made as pointed out in [15]:

- The only way to decrypt an encrypted message is to know the appropriate key.

- An encrypted message does not reveal the key that was used to encrypt it.

- The decryption algorithm used can always detect whether a message was encrypted with the expected key (i.e., there is adequate redundancy in the message).

5.6.1.1.5 Knowledge Unit   A knowledge unit $K_i \in K_1, K_2, \ldots, K_n$ consists of a knowledge base (which is composed of rules) and a set of facts. A knowledge unit reacts to new facts added to its fact list. K^i denotes the set of knowledge units belonging to process P_i. Knowledge units provide a level of intelligence for agents. This intelligence can be used to select what context a consumer requires, what context a consumer must generate, and so on.

K	::=	0	Empty unit
	|	r	Rule
	|	$K_1 + K_2$	Summation of knowledge units

The empty knowledge unit—one with no rules or facts—is denoted by 0. A knowledge unit may consist of a single rule. The summation $K_1 + K_2$ indicates that both knowledge units K_1 and K_2 react to a fact simultaneously, essentially behaving as a single knowledge unit.

5.6.1.1.6 Milieu   A milieu is a bounded place (also called an *environment*) in which processes reside and computations take place. Although milieus are very similar to ambients in ambient calculus (see [28] and [25]), Rahimi takes great care to differentiate the two: milieus are not a basic unit of a system, but rather represent an environment in which processes can join together to form a new computational unit [34]. Furthermore, the existence of separate locations is represented by a topology of these boundaries.

A milieu is surrounded by a border that must be traversed in order to join or leave it. We will show that communication can indeed occur through the milieu boundary via an unnamed environment channel (via the *listener* process). An entire milieu can move, taking with it its entire contents (i.e., all the processes and other milieus within it). Milieus are well suited to address the characteristics of the natural grouping and security of the system.

M	::=	0	Empty milieu
	$\mid$	$\beta.M$	Action prefix
	$\mid$	$M[O]$	Ownership
	$\mid$	$M[O_1 \mid O_2]$	Parallel
	$\mid$	$M_1 + M_2$	Summation of milieus

An empty milieu is denoted as 0. The variables $O_1, O_2, \ldots, O_n$ are used to range over processes and milieus. $M[O]$ is a milieu in which process or milieu O exists. A milieu may consist of other milieus or processes behaving in parallel; for example, $M[O_1 \mid O_2]$. The expression $M_1 + M_2$ indicates that milieu M is generated by the merging of milieus M_1 and M_2. The prefix β is an action prefix. The expression $\beta. M$ performs the action β and then behaves as M.

$M[O]$ exhibits a tree structure induced by processes and the nesting of milieu brackets (e.g., $M[P_1 \mid \ldots \mid P_p \mid M_1[\ldots] \mid \ldots \mid M_q[\ldots]]$). In API-S, process mobility is represented as the crossing of milieu boundaries; however, interactions between processes can cross the milieu boundary.

5.6.1.2 Actions
API-S inherits the traditional Send and Receive actions as defined in the π-calculus. Furthermore, knowledge unit and milieu actions are inherited from the API calculus. As such, terms and processes may be present

within actions. Knowledge unit actions include the receiving and sending of knowledge units and the adding and dropping of facts and rules. Milieu actions include the ability to join or leave a milieu, and the ability to open a milieu boundary. The process action prefix is denoted by α; the milieu action prefix is denoted by β.

Let A be the set of all α-actions in the calculus:

- τ is a silent, internal action.

- $x(\vec{L})$ is an input prefix where x is the input port or channel of a process which contains it and $\vec{L}$ is a tuple of processes or terms. The process $x(\vec{L}).P$ inputs an arbitrary number of terms or processes $\vec{L}$ on the channel x and then behaves as $P\{\vec{L}_1/\vec{L}\}$. All free occurrences of the names $\vec{L}$ in P are bound by the input action prefix $x(\vec{L})$ in P. A consumer C receiving contextual information i on a communication channel x could be modeled as $x(i).C$.

- $\overline{x}\vec{L}$ is an output prefix where x is the output port or channel of a process which contains it and $\vec{L}$ is a tuple of processes or terms. The process $\overline{x}\vec{L}.P$ outputs an arbitrary number of terms or processes $\vec{L}$ on channel x and then behaves as P. A producer P sending contextual information i to a consumer on a communication channel x could be modeled as $\overline{x}(i).P$.

- $(\vec{K})P$ makes the tuple of knowledge unit names $\vec{K}$ local to P.

- $K_i\langle\vec{a}\rangle(\vec{R})$ is a knowledge unit call. The expression $K_i\langle\vec{a}\rangle(\vec{R}).P$ calls the knowledge unit K_i, passing a list of facts $\vec{a}$. The result of this call is placed in $\vec{R}$. All free occurrences of $\vec{R}$ in P are bound by the prefix $K_i\langle\vec{a}\rangle(\vec{R})$ in P. In hyperdistribution, this could be used to model a consumer increasing its knowledge by gathering contextual information as it migrates from market to market.

- $x(\vec{K})$ is an input prefix where $\vec{K}$ is a knowledge unit that is received from a process via channel x. The expression $x(K_1).P$ receives a knowledge unit K_1 on channel x and then behaves as P. In hyperdistribution, it may be necessary to transfer one's knowledge base to another. This action provides support for receiving an entire knowledge unit.

- $\overline{x}\vec{K}$ is an output prefix where $\vec{K}$ is a knowledge unit that is sent by a process via channel x. The expression $\overline{x}\vec{K}.P$ sends a knowledge unit

K_1 on channel x and then behaves as P. We could model an entire consumer's knowledge base being sent to another consumer using this input prefix.

- $K_1(\vec{a})$ adds the tuple $\vec{a}$ to the fact list of K_i (if $\vec{a}$ is a tuple of facts) or to the rule base of K_i (if $\vec{a}$ is a tuple of rules). The expression $K_1(\vec{a}).P$ adds a to the fact list or rule base of K_i depending on the definition of a (whether it is a rule or a fact).

- $\overline{K_i}\vec{a}$ drops the tuple $\vec{a}$ from the fact list of K_i (if $\vec{a}$ is a tuple of facts) or from the rule base of K_i (if $\vec{a}$ is a tuple of rules). The expression $\overline{K_i}\vec{a}.P$ drops a from the fact list or rule base of K_i depending on the definition of a (whether it is a rule or a fact).

- *join* $m.P$ allows process P to join milieu m and then behave as P inside of m. In hyperdistribution, this could model a consumer joining a market.

- *leave* $m.P$ allows process P to leave milieu m and then behave as P outside of m. A producer leaving a market could be modeled in this way.

- $\langle \vec{L} \rangle$ is a broadcast output prefix such that the tuple of processes or terms $\vec{L}$ is broadcast to the surrounding milieu. Any processes within the milieu in which $\vec{L}$ was broadcast may receive it (see the next list item). There is no notion of a channel; one may think of the milieu as an environment channel, and processes defined to be listeners of—or associated with—a milieu can listen or send messages on this channel. If there is more than one process listening, one is chosen in a nondeterministic manner; naturally, the diffusion can be implemented in such a way that if there are several processes listening, then all of them receive the message. This action provides support for general broadcasts; in hyperdistribution, for example, a producer may want to blast contextual information to every consumer at a market.

- $(\vec{L})$ is a broadcast input prefix such that the tuple of processes or terms $\vec{L}$ is received upon the condition that P is within the milieu that initially broadcast $\vec{L}$ or that P is listening to the milieu. All free occurrences of the names $\vec{L}$ in P are bound by the input listening prefix $(\vec{L})$ in P. This action could model multiple consumers receiving contextual information naïvely blasted throughout a market.

Let B be the set of all β-actions in the calculus:

- *join m.M* indicates the case that milieu M joins milieu m and then behaves as M inside of m.

- *leave m.M* indicates that milieu M leaves milieu m and then behaves as M outside of m.

- *open.M* indicates that the boundary surrounding milieu M is dissolved; M ceases to exist. Any processes or other milieus that were inside of M behave as if they do not belong to M.

These actions assist in modeling groups of cooperating agents within larger groups. For example, a large group of producers and consumers may naturally divide into several cliques, and we may wish to model this behavior.

5.6.1.3 Binding

Consider the expressions $\bar{x}T.P$ and $x(T).P$. In each case, T (ranging over terms) is binding with the scope of P. An occurrence of a term within a process is bound if it is—or lies within the scope of—a binding occurrence of the term. An occurrence of a name within a process is free if it is not bound. We write $ft(P)$ for the set of terms that have a free occurrence in P. For example,

$$ft((\bar{z}y.0 + \bar{w}T.0)|\bar{x}u.0) = \{z, y, w, T, x, u\}$$

and

$$ft(vx((x(z).\bar{z}y.0 + \bar{w}v.0)|vu\bar{x}u.0)) = \{y, w, v\}$$

The free terms of a process limit its capabilities. Consider a process P and a term R. In order for P to send or receive R, to send via R, or to receive via R, it must be that $R \in ft(P)$. Therefore, a name must occur free in two processes if such a name can be used for communication between the processes. Of course, one process must provide a sending capability; the other must provide a receiving capability.

5.6.1.4 Substitution and Convertibility

A substitution is a function from names to names. This simply means that we substitute a received message for one that is represented in our expression. For example, an agent may send a message x over some communication channel we share; in our (message-receiving) expression, perhaps we have labeled this message y. In this case, y is simply a placeholder. Substitution means that we will substitute the received message x for our placeholder y. We write $\{x/y\}$ for the substitution that maps y to x. We can reword this in the following manner: we wish to substitute the name y present in some process P with the name x. In general, $\{x_1, x_2, \ldots, x_n / y_1, y_2, \ldots, y_n\}$ (where the y_i are pairwise distinct) maps each y_i to x_i. We use σ to range over substitutions. For a process P, the process of applying a substitution σ replaces each term x by $\sigma(x)$. In order to avoid the unintended capture of variables, we implement alpha-conversion whenever needed. For example, the result of $(y(z).\bar{z}x.0)\{z/x\}$ is $y(w).\bar{w}z.0$ for some name $w \neq z$, and the result of $(a(x).(vb)\bar{x}b.\bar{c}y.0)\{xb/yc\}$ is $a(z).(vd)\bar{z}d.\bar{b}x.0$ for some name $z \neq x$ and $d \neq b$. The calculus allows us to model the passing of actual communication channels. In hyperdistribution, this might allow us to establish secure channels, for example.

For convertibility, alpha-conversion is defined as follows:

- If the name w does not occur in the process P, then $P\{w/z\}$ is the process obtained by replacing each free occurrence of z in P by w.

- In process P, the replacement of $x(T).Q$ by $(x(R).Q)\{R/T\}$ is a change of bound terms where, in each case, R does not occur in Q.

- Processes P and Q are convertible (i.e., $P = Q$) if Q can be obtained from P by a finite number of changes of bound names.

For example:

$$(y(w).\bar{w}x.0)\{z/x\} = y(w).\bar{w}z.0$$

and

$$(!vz\bar{x}z.0 | y(w).0)\{v/x, v/y\} = !vz\bar{v}z.0 | v(w).0$$

5.6.1.5 Broadcasting

We make use of the listener process combined with the broadcast prefixes $\langle \bar{L} \rangle$ and $(\bar{L})$ so that we enable communication across the milieu boundary. We may, for example, associate a process with several milieus without the limitation that it needs to be inside those milieus; this enables the process to broadcast messages to processes within its associated milieus. Indeed, it may also listen to broadcasted messages originating from the same milieus; for example,

$$M_1[P|Q|\langle\langle R\rangle\rangle M_2[S]$$

Note that R is executing in parallel (i.e., concurrently with M_1, M_2, P, Q, and S): $\langle\langle R\rangle\rangle$ possesses an implied $|R|$. In this case, process R is *listening* to (associated with) milieus M_1 and M_2. Process P may broadcast a message x that process R can receive, and vice versa; for example,

$$M_1[\langle x\rangle.P|Q|\langle\langle (y).R\rangle\rangle M_2[S]$$

It is clear, in this case, that process R will receive message x as broadcasted by process P. The following, however, is ambiguous:

$$M_1[\langle x\rangle.P|Q|\langle\langle (y).R\rangle\rangle M_2[\langle z\rangle.S]$$

Which message is process R receiving? Indeed, it may very well be x as broadcasted by process P; however, since broadcasts are chosen in a non-deterministic manner, R may receive z from S instead. In order to remove this ambiguity, we allow associated processes (those listening to milieus and outside of them) to specify the environment channel when broadcasting or receiving a message; this is very similar to the standard input and output prefix syntax. For clarity, we use the milieu name as the channel name. For example,

$$M_1[\langle x\rangle.P|Q|\langle\langle M_1(y_1).M_2(y_2)R\rangle\rangle M_2[\langle z\rangle.S]$$

Now it is clear that process R will first receive x from P; it will subsequently receive z from S. We may, of course, omit the specification of the milieu if the model can capably tolerate it (i.e., we do not care if all of the processes within listening milieus can receive broadcasts).

Technically, this *environment channel* does not directly imply the true notion of a channel. It is simply a way to associate a process (or agent) to several milieus (or groups), thereby allowing it to be a part of several communication groups simultaneously. This behavior is more true to the real-world distributed systems we implement. Reconfigurable systems (and MAS) can be better modeled if we capably support such dynamics. Within the context of the hyperdistribution environment, this powerful abstraction allows us to model communication to numerous disjoint cooperative groups of consumers and producers.

5.6.1.6 Abbreviations

Quite often, we wish to reduce the size of the expressions we write in API-S. Typically, we apply *syntactic sugar*, which is simply a way of allowing us to abbreviate certain expressions:

- Sometimes, no message is communicated over a channel. To model this, we introduce a special name ε which is never bound; we may then write

$$\bar{x}.P \text{ in place of } \bar{x}\varepsilon.P$$

  and

$$x.P \text{ in place of } x(y).P, y \notin ft(P)$$

- We will often omit the *no action* process 0 as it is redundant. For example, we generally write

$$\bar{x}y \text{ in place of } \bar{x}y.0$$

- Often, we wish to allow input names to dynamically determine the course of computation and write

$$x(v).([v = y_1]P_1 + [v = y_2]P_2 + \ldots)$$

  where the y_i are distinct. Assuming that $v \notin ft(P_i)$, we reduce this to

$$x : [y_1 \Rightarrow P_1, y_2 \Rightarrow P_2, \ldots]$$

- We often abbreviate the successor term $suc(x)$ to its integer value:

$$suc(0) \Rightarrow 1$$

$$suc(suc(0)) \Rightarrow 2$$

- We define some composite prefixes:

$$\overline{x}y_1.y_2\ldots y_n \Rightarrow \overline{x}y_1.\overline{x}y_2\ldots\overline{x}y_n$$

$$x(y_1)(y_2)\ldots(y_n) \Rightarrow x(y_1).x(y_2)\ldots x(y_n)$$

$$join\ M_1 M_2\ldots M_n \Rightarrow join\ M_1.join\ M_2\ldots join\ M_n$$

$$leave\ M_1 M_2\ldots M_n \Rightarrow leave\ M_1.leave\ M_2\ldots leave\ M_n$$

- If a process must leave a milieu simply to communicate with another process (and then reenters the milieu),

$$l.\overline{x}a.j.P \Rightarrow leave\ M_1.\overline{x}a.join\ M_1.P$$

  Of course, the same applies to the reverse condition that a process must join a milieu in order to communicate with another process (and subsequently leaves the milieu):

$$j.x(a).l.P \Rightarrow join\ M_1.x(a).leave\ M_1.P$$

- If a process needs to leave an arbitrary n milieus in order to communicate with another process, then

$$l^n.\overline{x}a.j^n.P \Rightarrow leave\ M_1.leave\ M_2\ldots leave\ M_n.\overline{x}a.join\ M_n.join\ M_{n-1}\ldots$$

$$join\ M_1.P$$

- We utilize pair splitting when receiving a pair of terms or processes and when decrypting:

$$x(y_1,y_2).P \Rightarrow x(y).let\ (y_1,y_2) = y\ in\ P$$

$$case\ L\ of\ \{y_1,y_2\}_N\ in\ P \Rightarrow case\ L\ of\ \{y\}_N\ in\ let\ (y_1,y_2) = y\ inP$$

  We generalize pair splitting in the following manner:

$$(y_1,y_2,\ldots,y_n) \Rightarrow ((y_1,y_2,\ldots,y_{n-1}),y_n)$$

$$let\ (y_1,y_2,\ldots,y_n) = N\ in\ P \Rightarrow let\ (y,y_n) = N\ in\ let\ (y_1,y_2,\ldots,y_{n-1}) = y\ in\ P$$

- If a listener process is associated with a single milieu, we write

$$M[\,]\langle\!\langle P \text{ in place of } M[\,]\langle\!\langle P \rangle\!\rangle$$

$$M_1[\,]\langle\!\langle P\,|\,M_2[\,] \text{ in place of } M_1[\,]\langle\!\langle P \rangle\!\rangle|M_2[\,]$$

The syntax of the API-S often alludes to things similar in nature. Take, for example, the following set of expressions:

$$x(m_1).\bar{x}m_1 \quad \text{and} \quad x(m_2).\bar{x}m_2$$

Both intuitively convey the same thing, in that one process inputs some term or process along a channel and another outputs some term or process along a channel. This is a simple synchronization of processes. The ambiguity of the messages m_1 and m_2 does not matter; both expressions exhibit the same behavior. Take another example:

$$P|Q \quad \text{and} \quad Q|P$$

Again, both exhibit the same behavior: the parallel composition of processes P and Q. In order to identify the processes and expressions which intuitively exhibit the same behavior and thus represent the same thing, we introduce a *structural congruence*.

5.6.1.7 Structural Congruence

The structural congruence $\equiv$ is defined as the congruence on processes that satisfies the laws outlined in Table 5.3. Note that if processes P and Q are variants of the alpha-conversion, then $P \equiv Q$. Furthermore, if $P \equiv Q$ can be inferred from the structural axioms and equational-reasoning rules, then P and Q are structurally congruent. Take the following example [38]:

$$P = vx(x(z).\bar{z}a.0|(\bar{x}y.\tau|\bar{x}b.0))$$

By applying the axioms of associativity and commutativity of composition coupled with the identity of τ, we obtain that

$$P \equiv P_1 = vx(((\bar{x}y.0+0)|(x(z).\bar{z}a.0+0))|\bar{x}b.0)$$

TABLE 5.3   Axioms of Structural Congruence and Rules of Equational Reasoning in API-S

Structural Congruence	
$[T = T]\tau.P \equiv \tau.P$	Match
$P + (Q + R) \equiv (P + Q) + R$	Summation associativity
$P + Q \equiv Q + P$	Summation commutativity
$P + 0 \equiv P$	Summation identity
$P\|(Q\|R) \equiv (P\|Q)\|R$	Composition associativity
$P + P \equiv P$	Same process
$P\|Q \equiv Q\|P$	Composition commutativity
$P\|0 \equiv 0$	Composition identity
$vT_1 vT_2 P \equiv vT_2 vT_1 P$	Restriction
$vT0 \equiv 0$	Restriction identity
$vT(P_1\|P_2) \equiv P_1\|vTP_2$, if $T \notin ft(P_1)$	Restriction composition
$!P \equiv P\|!P$	Replication
Equational Reasoning	
$P = P$	Reflexivity
$P = Q \Rightarrow Q = P$	Symmetry
$P = Q \wedge Q = R \Rightarrow P = R$	Transitivity
$P = Q \Rightarrow C[P] = C[Q]$	Generality

It should be noted that in the transformation from P to P_1, two potential interactors have been brought together. We may write, equally, that

$$P \equiv P_2 = vx(((\bar{x}b.0 + 0)\|(x(z).\bar{z}a.0 + 0))\|\bar{x}y.0)$$

which implies a change of potential interactors.

5.6.1.8 Reduction

The reduction relation $\rightarrow$ is defined on processes and milieus according to the rules outlined in Table 5.4. It merely provides a mechanism whereby some process P can evolve to P' as a result of some interaction (an action within P). Reduction is inferred directly from the syntax and is defined by

TABLE 5.4   Reduction Rules in API-S

TAU	$\tau.P + q \rightarrow P$
REACT	$(x(\vec{T}).P + P') \| (\overline{x}\vec{R}.Q + Q') \rightarrow P\{\vec{R}/\vec{T}\} \| Q$
Parallel inference rule, PAR	$\dfrac{P \rightarrow P'}{P\|Q \rightarrow P'\|Q}$
Restriction inference rule, RES	$\dfrac{P \rightarrow P'}{(vx)P \rightarrow (vx)P'}$
RES-K	$\dfrac{P \rightarrow P'}{(K)P \rightarrow (K)P'}$
MIL	$\dfrac{P \rightarrow P'}{M[P] \rightarrow M[P']}$
STRUCT	$\dfrac{P \equiv P', P \rightarrow Q, Q \equiv Q'}{P' \rightarrow Q'}$

a family of inference rules [19]. We begin by stating the communication axiom:

$$(\cdots + \overline{x}L_1.P_1) \| (\cdots + x(L_2).P_2) \rightarrow P_1 | P_2\{L_1/L_2\} \tag{5.1}$$

The process prior to the reduction consists of two main parallel processes: a sending process and a receiving process. The term or process L_1 is sent along channel x, and L_2 is received via channel x. The axiom expresses that P has a reduction arising from an interaction between its components via channel or port x. As a result of this reduction, L_1 is *passed* from the first process to the second and is substituted for the placeholder L_2 in P_2; the two prefixes are consumed.

A second axiom is used in defining reduction:

$$\tau.P \rightarrow P \tag{5.2}$$

where the τ prefix expresses an internal action whose origin is not necessarily explicit.

For closed processes, the following reductions hold:

$let\ (x, y) = (M, N)\ in\ P$	$\rightarrow$	$P\{MN/xy\}$
$case\ 0\ of\ 0 : P, suc(x) : Q$	$\rightarrow$	P
$case\ suc(M)\ of\ 0 : P, suc(x) : Q$	$\rightarrow$	$P\{M/x\}$
$case\ \{M\}_N\ of\ \{x\}_N\ in\ P$	$\rightarrow$	$P\{M/x\}$
$case\ \{[M]\}_{N+}\ of\ \{[x]\}_{N-}\ in\ P$	$\rightarrow$	$P\{M/x\}$
$case\ [\{M\}]_{N-}\ of\ [\{x\}]_{N+}\ in\ P$	$\rightarrow$	$P\{M/x\}$

We now introduce the structural rule:

$$\text{STRUCT:}\quad \frac{P \equiv P', P \rightarrow Q, Q \equiv Q'}{P' \rightarrow Q'} \tag{5.3}$$

which is read as follows:

From $P \equiv P'$ and $P \rightarrow Q$ and $Q \equiv Q'$, infer $P' \rightarrow Q'$

Note that the structural rule can be rewritten as follows to mean the same thing:

$$\text{STRUCT:}\quad \frac{P \equiv Q, P \rightarrow P', P' \equiv Q'}{Q \rightarrow Q'}$$

The inference rules for reduction are defined as follows:

$$\text{PAR:}\quad \frac{P \rightarrow P'}{P|Q \rightarrow P'|Q} \tag{5.4}$$

$$\text{RES:}\quad \frac{P \rightarrow P'}{(vx)P \rightarrow (vx)P'} \tag{5.5}$$

In the PAR, we assert that if the component P of the process $P|Q$ has a reduction, then $P|Q$ has a reduction, the effect of which is just the effect on P. The component Q is unaffected by the action within P. The RES simply expresses that a restriction in some term does not inhibit reduction in a process.

The API-S calculus includes a new inference rule which directly addresses knowledge unit restriction:

$$\text{RES-K: } \frac{P \to P'}{(K)P \to (K)P'} \tag{5.6}$$

The two axioms defined in Equations 5.1 and 5.2, the structural rule defined in Equation 5.3, and the inference rules defined in Equations 5.4, 5.5, and 5.6 collectively define *reduction*. For an example, consider the following expression:

$$\bar{x}y.0 | vy(x(z).Q\{y/z\})$$

We may reduce this expression using the reduction rules and structural congruence as follows:

	$\bar{x}y.0 | vy(x(z).Q\{y/z\})$	
$\equiv$	$\bar{x}y.0 | vy'(x(z).Q\{y'/z\})$	Alpha-conversion
$\equiv$	$vy'(\bar{x}y.0 | x(z).Q\{y'/z\})$	Restriction composition
$\to$	$vy'(0 | Q\{y'/y\})$	REACT
$\equiv$	$vy'(Q\{y'/y\})$	Composition identity

In another example, the expression

$$vx((x(u).Q_1 + y(v).Q_2) | \bar{x}a.0) | (\bar{y}b.R_1 + \bar{x}c.R_2)$$

may also be reduced using the reduction rules. First, we isolate a part of the expression:

$$(x(u).Q_1 + y(v).Q_2) | \bar{x}a.0$$

and note that it matches the left side of the REACT rule. We now have that

$(x(u).Q_1 + y(v).Q_2) | \bar{x}a.0 \to Q_1 | 0$	REACT
$(x(u).Q_1 + y(v).Q_2) | \bar{x}a.0 \to Q_1$	STRUCT
$vx((x(u).Q_1 + y(v).Q_2) | \bar{x}a.0) \to vx(Q_1)$	RES

and can now reduce the complete expression as follows (via PAR):

$$vx((x(u).Q_1 + y(v).Q_2)|\bar{x}a.0)|(\bar{y}b.R_1 + \bar{x}c.R_2) \rightarrow vx(Q_1)|(\bar{y}b.R_1 + \bar{x}c.R_2)$$

5.6.1.9 Simple Examples of API-S

In this section, our aim is to illustrate the flexibility of API-S. In order to more fully illustrate the power of the calculus (e.g., the natural grouping ability that milieus offer), we provide several examples that offer a view of some of the core components of the calculus. For more examples relating to the nonsecurity aspects of the calculus, the reader is referred to [34].

5.6.1.9.1 Passing Links: The Client–Server–Printer Example   Server S shares a link x with printer P; client C also shares a link y with the server. C wishes to send a document d to P for printing. In this example, we illustrate the passing of the communication link x from S to C so that the client can use this link to send its document. The act of printing the document is modeled by the function $F(\ )$, an internal function in P. We define the system illustrated in Figure 5.10 as follows:

$$S = \bar{y}x.S'$$
$$P = x(d').F(d').P'$$
$$C = y(x').\bar{x}'d.C'$$

The behavior is then

$$\bar{y}x.S'|x(d').F(d').P'|y(x').\bar{x}'d.C'$$
$$\rightarrow \quad S'|x(d').F(d').P'|(\bar{x}'d.C')\{x/x'\}$$
$$\rightarrow \quad S'|(F(d').P')\{d/d'\}|C'$$
$$\rightarrow \quad S'|P'|C'$$

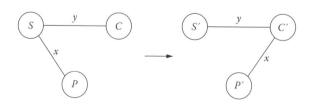

FIGURE 5.10   Passing links: the client–server–printer example.

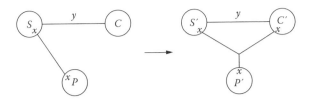

FIGURE 5.11 Passing links with restriction.

Note that the communication links x and y are indeed global (i.e., not private to any particular process). If, for example, we restrict x to processes S and P initially, we obtain a much different result (as illustrated in Figure 5.11). Indeed, the behavior is very similar:

$$vx(\bar{y}x.S'|x(d').F(d').P')|y(x').\bar{x}'d.C'$$

$$\rightarrow \quad vx(S'|x(d').F(d').P'|(\bar{x}'d.C')\{x/x'\})$$

$$\rightarrow \quad vx(S'|(F(d').P')\{d/d'\}|C')$$

$$\rightarrow \quad vx(S'|P'|C')$$

When communication link x is passed from S to C, its scope *extrudes* and is extended. It would be an entirely different scenario if, for example, client C possessed a public channel x initially. In this case, we rename the passed channel such that, again, its scope is extruded, and maintain the name of the client's public channel x.

5.6.1.10 Examples with Knowledge Units   Often, we wish to model the secret exchange (or just simple transmission) of a complete knowledge unit or perhaps even a simple fact or rule between agents. The following examples illustrate the use of knowledge units coupled with simple cryptographic protocols. We could similarly model the dissemination of contextual information from producer to consumer in this way.

5.6.1.10.1 Passing a Knowledge Unit Using Shared Key Cryptography   Consider two agents A_1 and A_2; they share a link x. Agent A_1 possesses a knowledge unit K_1; agent A_2 possesses a knowledge unit K_2. Agent A_1 wishes to *hand off* its knowledge unit K_1 to agent A_2 using a private key S they both share. This example (illustrated in Figure 5.12) can be simply modeled in the calculus as follows:

$$A_1 = \bar{x}\{K_1\}_S.A_1'$$
$$A_2 = x(K).case\ K\ of\ \{y\}_S\ in\ A_2'$$

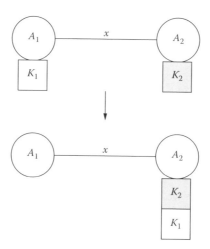

FIGURE 5.12   Passing a knowledge unit.

The behavior is then as follows:

$$\overline{x}\{K_1\}_S.A_1'|x(K).case\ K\ of\ \{y\}_S\ in\ A_2'$$
$$\rightarrow \quad A_1'|(case\ K\ of\ \{y\}_S\ in\ A_2')\{K_1/K\}$$
$$\rightarrow \quad A_1'|A_2'\{K_1/y\}$$

The result is such that agent A_2 now possesses knowledge units K_1 and K_2.

One may wish agent A_1 to simply pass a *copy* of its knowledge unit K_1 to A_2, thus maintaining ownership of K_1. In this case, we must alter the definitions of A_1 and A_2 appropriately:

$$A_1 = vK_1(\overline{x}\{K_1\}_S.A_1')$$
$$A_2 = x(K).case\ K\ of\ \{y\}_S\ in\ A_2'$$

Utilizing the restriction, the behavior now becomes (as expected):

$$vK_1(\overline{x}\{K_1\}_S.A_1')|x(K).case\ K\ of\ \{y\}_S\ in\ A_2'$$
$$\equiv \quad vK_1(\overline{x}\{K_1\}_S.A_1'|x(K).case\ K\ of\ \{y\}_S\ in\ A_2')$$
$$\rightarrow \quad vK_1(A_1'|(case\ K\ of\ \{y\}_S\ in\ A_2')\{K_1/K\})$$
$$\rightarrow \quad vK_1(A_1'|A_2'\{K_1/y\})$$

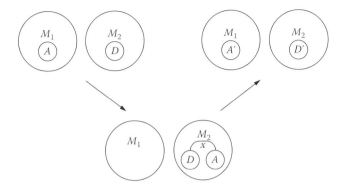

FIGURE 5.13 Consumer information request (with migration).

5.6.1.11 Examples with Milieus   Milieus present a unique way to model the natural grouping often seen in MAS. Intrinsically, they offer some rudimentary aspects of security (e.g., authentication), particularly when combined with the typing and naming rules of the calculus. Furthermore, they are very well suited to model the agency which forms an integral part of the mobile agent paradigm.

5.6.1.11.1 Migrating Consumer with Information Request   We wish to illustrate that API-S can capably model a real-world example, albeit a fairly simple one. In the scenario illustrated in Figure 5.13, there exist two markets M_1 and M_2 (modeled as milieus), one consumer A, and one database D (a process). A, initially at market M_1, will migrate to M_2, query D with a request for contextual information q, receive a response r from D, and return home. A simple way to implement this scenario in the calculus is as follows:

$$M_1 = [A]$$
$$M_2 = [D]$$
$$A = leave\ M_1.join\ M_2.\bar{x}q.x(r').leave\ M_2.join\ M_1.A'$$
$$D = x(q').\bar{x}r.D'$$

The behavior of the system becomes

$$M_1[leave\ M_1.join\ M_2.\bar{x}q.x(r').leave\ M_2.join\ M_1.A']\,|\,M_2[x(q').\bar{x}r.D']$$
$$\rightarrow\quad M_1[\ ]\|M_2[\bar{x}q.x(r').leave\ M_2.join\ M_1.A'|x(q').\bar{x}r.D']$$
$$\rightarrow\quad M_1[\ ]\|M_2[x(r').leave\ M_2.join\ M_1.A'|(xr.D')\{q/q'\}]$$
$$\rightarrow\quad M_1[\ ]\|M_2[(leave\ M_2.join\ M_1.A')\{\bar{x}/r'\}|D']$$
$$\rightarrow\quad M_1[A']\|M_2[D']$$

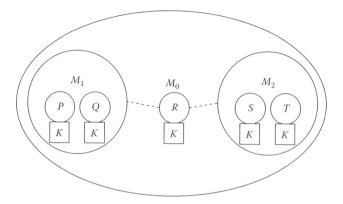

FIGURE 5.14   Communicating through the milieu boundary via broadcast.

5.6.1.11.2 Communicating through the Milieu Boundary via Broadcast   By combining cryptographic primitives with the listener, we show that it is possible to communicate through the milieu boundary quite securely. In this example, we wish to model the system as illustrated in Figure 5.14. In this scenario, agent P would like to send a fact a to agent S. A shared key k known to all agents is used to encrypt and decrypt a. The agent may indeed be a producer communicating an encrypted bit of contextual information to a consumer. To accomplish this, we will use agent R as an intermediary. We define the system and scenario as follows:

$$M_0 = [M_1 \langle\!\langle R \rangle\!\rangle M_2]$$
$$M_1 = [P|Q]$$
$$M_2 = [S|T]$$
$$P = \langle \{a\}_k \rangle.P'$$
$$R = (b).\langle b \rangle.R$$
$$S = (c).case\ c\ of\ \{d\}_k\ in\ K(d).S'$$

The behavior is then

$$M_0[M_1[\langle \{a\}_k \rangle.P'|Q]\langle\!\langle (b).\langle b \rangle.R' \rangle\!\rangle M_2[(c).case\ c\ of\ \{d\}_k\ in\ K(d).S'|T]]$$
$$\rightarrow \quad M_0[M_1[P'|Q]\langle\!\langle (\langle b \rangle.R')\{a/b\} \rangle\!\rangle M_2[(c).case\ c\ of\ \{d\}_k\ in\ K(d).S'|T]]$$
$$\rightarrow \quad M_0[M_1[P'|Q]\langle\!\langle R' \rangle\!\rangle M_2[(case\ c\ of\ \{d\}_k\ in\ K(d).S')\{a/c\}|T]]$$
$$\rightarrow \quad M_0[M_1[P'|Q]\langle\!\langle R' \rangle\!\rangle M_2[(K(d).S')\{a/d\}|T]]$$
$$\rightarrow \quad M_0[M_1[P'|Q]\langle\!\langle R' \rangle\!\rangle M_2[S'|T]]$$

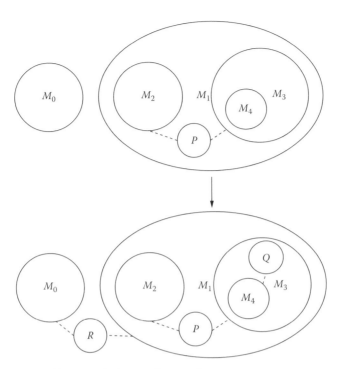

FIGURE 5.15 Milieu association workaround.

Naturally, a process may only be associated with milieus at its own level (i.e., only those that it can directly join) or one that directly surrounds it. However, we can simulate the association of unassociable milieus by placing a series of processes as intermediaries in such a way that association is ultimately derived. For example, in Figure 5.15, process P is associated with milieus M_1, M_2, and M_3. If association with milieu M_4 is desired, we simply introduce a process Q in M_3 and associate it with M_4; Q is now the intermediary between any process directly in M_4 and any process directly in M_1. Similarly, if association with milieu M_0 is desired, we introduce a process R outside of M_1 and associate it with both M_0 and M_1; R is now the intermediary between any process directly in M_0 and any process directly in M_1.

We should note that there may indeed be more than one process or agent associated with the same milieu(s); for example, the following is a valid expression in the calculus:

$$M_1[P|Q]\langle\langle R|S\rangle\rangle M_2[T]$$

5.7 EXAMPLE: CONTEXTUAL HYPERDISTRIBUTION

Consider the tsunami scenario. In a typical hierarchical IT system, the data corresponding to the earthquake generated by earthquake sensors, the subsequent wave measurements in the ocean picked up by numerous buoys, and any other relevant data would be obtained from a few potential disparate agencies with access to the data. In many cases, the agencies may be unaware that the others exist or, more importantly, that their data may in fact be correlated. The data would then be routed to another entity whose task it is to make a tsunami prediction. If such an event were probable, a select number of individuals would be alerted; and the process would continue—*down the line*—across the globe until the general public is informed, often too late. At any point in this tedious process, a failure could occur, thereby halting the dissemination of information. In order to route information from one component to another, knowledge of the routing entities must be known; this is a major problem of current IT systems. Moreover, because of the fragmentation of information caused by this scenario and the lack of routing entities, there can be limited dissemination of information about the event when in fact consumers of information might be almost any person or entity on the shores of the Indian Ocean, for example. The problem of the distribution of information to entities that might be interested in such information is a very real problem in current IT systems.

On the other hand, hyperdistribution provides a way of disseminating information to those who need it (people in danger of the effects of a probable tsunami) without presenting any significant points of failure. Support for mapping-related concepts is provided via the ontology. The redundancy, robustness, and scalability of the *flat* hyperdistribution model preclude global failure of information dissemination. Moreover, decisions can be made at the peer level so that, for example, a tsunami prediction could be made and verified dynamically by a group of acquainted consumers. Propagation of time-critical information occurs quickly as many producers can disseminate it to other consumers within their locality. These consumers then become producers and subsequently repeat the process.

In a traditional IT system, the process of discovering consumers that might be interested in a type of thematic information provided by a producer relies on communicating through a single entity (or relatively few entities). Consumers must be itemized in some manner and known by producers a priori. Dynamic discovery of consumers is nonexistent. Hyperdistribution allows for the discovery of consumers of information mathematically by the principle of *degrees of separation* as in social networks. The basis for discovery is the *knows relationship*.

Not only can producers disseminate information directly to consumers within the hyperdistribution environment, but also markets can be used as intermediaries, temporarily storing contextual information destined for needy consumers. This is something that is nonexistent in traditional IT systems. The overall dynamic and nondeterministic nature of hyperdistribution creates an environment in which information just flows naturally.

Hyperdistribution in the tsunami example may consist of consumers that are instantiated at the onset of some earthquake event within some predetermined geographical distance. Their initial bit of information consists of an earthquake event. It is now their task to determine if the earthquake is something to be worried about—and thus worrisome to the public they ultimately represent. To this end, they must obtain the surrounding context for the earthquake event. Suppose the earthquake originated somewhere in the ocean. The consumers may, for example, quickly obtain additional information related to reports of damaged buildings via producers. That information may eliminate the possibility of immediate danger. Concurrently, producers may be monitoring ocean buoys that pick up anomalies soon after the earthquake. The consumers need this contextual information (i.e., an earthquake has occurred in the ocean and nearby buoys have reported anomalies signifying a rise in ocean waves) in order to determine whether or not an evacuation alert must ultimately be made.

The tsunami event is similar to a drop of water falling into a lake. The drop is the source of the event, and its occurrence must be propagated to consumers. Like the ripples in the lake that are generated at the drop's contact, the information spreads to consumers throughout the system, as shown in Figure 5.16.

5.8 RESEARCH DIRECTIONS IN HYPERDISTRIBUTION OF CONTEXTS

Dealing with the uncertainty of data and contextual information (e.g., missing pieces of data that a consumer might need in order to decide whether to perform some action) is a problem that can be addressed via soft computing. Techniques such as fuzzy logic can be used to consider the case when a consumer lacks contextual information but must nonetheless take action. Indeed, there are other techniques that can be employed, and researching these is left for the future.

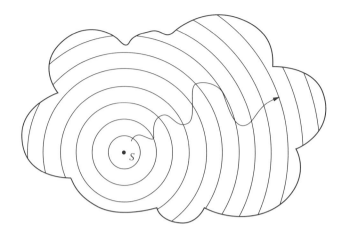

FIGURE 5.16   Context propagation in hyperdistribution (*s* is event source).

As presented, our agent-based hyperdistribution model supports various broad forms of security; for example, basic encryption of data and communication channels and other specific security techniques typically relevant to the agent paradigm. Various advanced topics have been discussed that provide powerful modeling tools to validate such systems with security requirements. In the future, it will be necessary to provide a formal model for security within the hyperdistribution of contextual information. API-S could certainly be utilized to form the beginnings of such a model.

There are numerous other areas of research relevant to this area; for example,

- The specification of principles for an ontological model for hyperdistribution

- The design of a producer and consumer ontology for use in determining and generating contextual information

- A more "intelligent" producer–consumer agent model that extends the rule-based design proposed in this chapter

- Investigation into security methods to assist in the dissemination of contextual information

The next chapter addresses one of the key architectural considerations in contextual processing: data management. Because contextual data will have complex relationships based on time and space, current methods for management of complex data, such as contextual information, can be limited. A brand-new paradigm for repository management, storage, and retrieval will be presented and discussed.

REFERENCES

1. Lange, D. B., and M. Oshima. (1999). Seven good reasons for mobile agents. *Communications of the ACM* 42 (3): 88–89.
2. Picco, G. P. (2001). Mobile agents: An introduction. *Journal of Microprocessors and Microsystems* 25 (2): 65–74.
3. Du, T. C., E. Y. Li, and A.-P. Chang. (2003). Mobile agents in distributed network management. *Communications of the ACM* 46 (7): 127–132.
4. Boughaci, D., K. Ider, and S. Yahiaoui. (2007). Design and implementation of a misused intrusion detection system using autonomous and mobile agents. In *EATIS '07: Proceedings of the 2007 Euro American Conference on Telematics and Information Systems*, pp. 1–8. New York: ACM Press.
5. Rahimi, S., and N. F. Carver. (2005). A multi-agent architecture for distributed domain-specific information integration. In *HICSS '05: Proceedings of the 38th Annual Hawaii International Conference on System Sciences (HICSS'05)—Track 4*, pp. 113.2. Washington, DC: IEEE Computer Society.
6. Wooldridge, M., and N. R. Jennings. (1995). Intelligent agents: Theory and practice. *Knowledge Engineering Review* 10 (2): 115–152.
7. Poole, D., A. Mackworth, and R. Goebel. (1997). *Computational intelligence: A logical approach*. Oxford: Oxford University Press.
8. Schoeman, M., and E. Cloete. (2003). Architectural components for the efficient design of mobile agent systems. In *SAICSIT '03: Proceedings of the 2003 Annual Research Conference of the South African Institute of Computer Scientists and Information Technologists on Enablement through Technology*, pp. 48–58. Port Elizabeth: South African Institute for Computer Scientists and Information Technologists.
9. Bellavista, P., A. Corradi, and S. Monti. (2005). Integrating web services and mobile agent systems. In *ICDCSW'05: Proceedings of the First International Workshop on Services and Infrastructure for the Ubiquitous and Mobile Internet (SIUMI)*, pp. 283–290. Washington, DC: IEEE Computer Society.
10. Chess, D., C. Harrison, and A. Kershenbaum. (1994). *Mobile agents: Are they a good idea?* (Technical Report RC 19887). Yorktown Heights, NY: IBM Research Division.
11. Chess, D. M. (1998). Security issues in mobile code systems. In *Mobile agents and security*, pp. 1–14. London: Springer-Verlag.
12. Kun, Y., G. Xin, and L. Dayou. (2000). Security in mobile agent system: Problems and approaches. *SIGOPS Operating Systems Review* 34 (1): 21–28.
13. Luck, M., P. McBurney, and C. Preist. (2004). A manifesto for agent technology: Toward next generation computing. *Autonomous Agents and Multi-Agent Systems* 9 (3): 203–252.

14. Schoeman, M. A. (2003). *Architectural guidelines for mobile agent systems* (Technical Report TR-UNISA-2003-02). Pretoria: School of Computing, University of South Africa.

15. Bhargavan, K., R. Corin, C. Fournet, and A. D. Gordon. (2004). Secure sessions for web services. In *SWS'04: Proceedings of the 2004 Workshop on Secure Web Services*, pp. 56–66. New York: ACM Press.

16. Bhargavan, K., C. Fournet, and A. Gordon. (2006). Verified reference implementations of WS-security protocols. Paper presented at the *3rd International Workshop on Web Services and Formal Methods* (*WS-FM 2006*), Vienna, September.

17. Bhargavan, K., C. Fournet, and A. D. Gordon. (2004). A semantics for web services authentication. In *POPL '04: Proceedings of the 31st ACM SIGPLAN-SIGACT Symposium on Principles of Programming Languages*, pp. 198–209. New York: ACM Press.

18. OASIS. (2004). WS-Reliability. http://docs.oasis-open.org/wsrm/ws-reliability/v1.1/wsrm-ws_reliability-1.1-spec-os.pdf

19. Parrow, J. (2001). An introduction to the pi-calculus. In J. Bergstra, A. Ponse, and S. Smolka (eds.), *Handbook of process algebra*, pp. 479–543. Amsterdam: Elsevier.

20. Moen, A. (2003). Introduction to pi-calculus. Paper presented at a seminar in computer science at the University of Oslo, March.

21. Milner, R., J. Parrow, and D. Walker. (1992). A calculus of mobile processes, part I. *Information and Computation* 100 (1): 1–40.

22. Milner, R., J. Parrow, and D. Walker. (1992). A calculus of mobile processes, part II. *Information and Computation* 100 (1): 41–77.

23. Wing, J. M. (2002). FAQ on pi-calculus (Microsoft internal memo). Redmond, WA, December.

24. Serugendo, G. D. M., M. Muhugusa, and C. F. Tschudin. (1998). A survey of theories for mobile agents. *World Wide Web* 1 (3): 139–153.

25. Gordon, A. D. (2000). Part III: The ambient calculus. Paper presented at the *International Summer School on Foundations of Security Analysis and Design* (*FOSAD 2000*), Bertinoro, Italy, September.

26. Cardelli, L., and A. D. Gordon. (1998). Mobile ambients. In *Foundations of Software Science and Computation Structures: First International Conference* (*FOSSACS '98*). Berlin: Springer-Verlag.

27. Cardelli, L., G. Ghelli, and A. D. Gordon. (1999). Mobility types for mobile ambients. In *ICAL '99: Proceedings of the 26th International Colloquium on Automata, Languages and Programming*, pp. 230–239. London: Springer-Verlag.

28. Cardelli, L., G. Ghelli, and A. D. Gordon. (2002). Types for the ambient calculus. *Information Computation* 177 (2): 160–194.

29. Desel, J., and G. Juhas. (2001). What is a Petri net? In *Unifying Petri nets: Advances in Petri nets*, pp. 1–25. London: Springer-Verlag.

30. Murata, T. (1989). Petri nets: Properties, analysis and applications. In *Proceedings of the IEEE*, pp. 541–580. New York: IEEE.

31. Peterson, J. L. (1977). Petri nets. *ACM Computing Surveys* 9 (3): 223–252.

32. Caselli, S., G. Conte, and P. Marenzoni. (2001). A distributed algorithm for GSPN reachability graph generation. *Journal of Parallel Distributed Computing* 61 (1): 79–95.

33. Rana, O. F. (2000). Performance management of mobile agent systems. In *AGENTS '00: Proceedings of the Fourth International Conference on Autonomous Agents*, pp. 148–155. New York: ACM Press.

34. Rahimi, S. (2002). API-calculus for intelligent-agent formal modeling and its application in distributed geospatial data conflation. PhD thesis, University of Southern Mississippi, Hattiesburg.

35. Sangiorgi, D. (1993). From pi-calculus to higher-order pi-calculus—and back. In *TAPSOFT '93: Proceedings of the International Joint Conference CAAP/FASE on Theory and Practice of Software Development*, pp. 151–166, London: Springer-Verlag.

36. Coppersmith, D. (1994). The data encryption standard (DES) and its strength against attacks. *IBM Journal of Research and Development* 38 (3): 243–250.

37. RSA Laboratories. (2001). *Pkcs#1 v2.1: RSA cryptography, standard draft 2.* Bedford, MA: RSA Laboratories.

38. Sangiorgi, D., and D. Walker. (2001). *Pi-calculus: A theory of mobile processes.* New York: Cambridge University Press.

CHAPTER 6

Set-Based Data Management Models for Contextual Data and Ambiguity in Selection

THEME

This chapter introduces the idiosyncrasies of managing context data, sets of context data related by theme, and supersets of context data. It develops a fuzzy metadata-based model for management of aggregated sets of data and proposes methods of dealing with the selection and retrieval of context data that are inherently ambiguous about what to retrieve for a given query.

This work builds on previous work and extends that work to incorporate the contextual model. It supports the previously defined contextual dimensions of time, space, similarity, and impact. The original model for spatial-temporal management is presented and discussed, and then a new model is derived from that particular model.

Additionally, the concept of *coverage* is introduced. Coverage has normally been thought of as data pertaining to a physical spatial region of the earth that is stored in a computer. Extending this definition, a context's coverage as discussed below is contextual information that applies to one of the dimensions of contexts, that of space, time, impact, or similarity.

As in previous chapters, this topic explains the issues of data management for context information and proposes a *potential framework* and model supporting the key components required to manage contextual data. It does not prescribe the specifics of implementation, only the components that might be included in the implementation. Thus, the chapter is meant to point in potential directions that could be pursued for research and further investigation.

6.1 INTRODUCTION TO DATA MANAGEMENT

Having examined in previous chapters how contexts are created, what they contain, and how they can be analyzed, it becomes clear that the data management issues of contexts are not readily solved by traditional approaches. Data management consists primarily of storage of information in a way that the relationships among the entities are preserved. Due to the fact that a context can really be composed of any type of data ranging from binary to images, narratives, or audio, there is a need for a new model for storage of context data that can handle widely different types of data. Additionally, contextual data management involves the issues of correlations by thematic relationship between related types of data. For example, context $C1$ may be very similar to context $C13–C21$ for a given thematic event; thus, they should be interrelated and included in the process of analysis and knowledge creation operations. Related to this idea is that similarity in contexts also is the driving force in the how and what of which contexts are retrieved for a given query.

With the above in mind, there is a need to develop a new model for data management of contexts. This model should present an architectural overview of how such a model might be created, what the elements in such a model are, and how it functions. The first part of the chapter presents an argument for a new type of paradigm of how data should be thought of, the *set model*. Problems with this new way of thinking about data organization are then discussed as a beginning for a discussion of solutions to the problems of sets. The chapter then continues on to discuss a new method of modeling sets that gives the architecture the ability, in theory, to store, manage, and retrieve any type of data currently existing and any type of data that may be created in the future. Finally the chapter presents concepts about how the coverage overlap problems found with the use of the set model can create ambiguity in retrieval and can be addressed with fuzzy set-based new operators that can identify similarities in data based on time, space, and contextual similarity and retrieve the best candidates to satisfy a given query.

6.2 BACKGROUND ON CONTEXTUAL DATA MANAGEMENT

Spatial information science is a relatively new and rapidly evolving field. Because global contextual models are highly spatial in many aspects of their operation, it is appropriate to look at the issues of context-based data management in terms of how spatial data are managed.

Spatial data management systems are an integration of software and hardware tools for the input, analysis, display, and output of spatial data and associated attributes. These systems are being used across a broad range of disciplines for analysis, modeling, prediction, and simulation of spatial phenomena and processes. Applications of spatial data are diverse: natural resource management, traffic control and road building, economic suitability, geophysical exploration, and global climate modeling are just a few examples. In the case of contextual data management, spatial data systems need to be extended for the purposes of managing wide types of information such as sensed images (e.g., aerial photos and satellite images) and to store data in a number of different raster and vector data structures. A contextual management system based on spatial data management principles may contain digital images, tabular survey data, and text, among many other possibilities.

Current spatial data management systems have limitations. One of these is an inability to retrieve and present all the data *relevant* to a given theme and then present such data to the system user in an orderly and helpful manner. When, for example, a user wants to access information about a particular geospatial region or type of geospatial feature (e.g., tsunami distance and travel information), he or she selects what appears to be an appropriate data entity. This process can be frustrating and error prone, because, typically, many data entities (maps, photos, etc.) contain information of that type, and choosing among them, or even being able to view them all, is very difficult. Furthermore, there may be *several data entities* (e.g., maps, photos, and sensor information) for the same thematic area that may have been collected and/or modified at different times for different reasons that makes selection of the correct data entity complicated for

TA about Diseased Trees

Satellite data on computer X   Textual description on computer Z   Database - map data

FIGURE 6.1   Distributed locations of information that collectively indicate a tsunami.

a given query. In addition, not all data related to a theme may be stored in a known location such as a given database. For instance, some relevant data may be in files, scattered across several computer systems' hard disks. As an example, say that a tsunami has just occurred in the Indian Ocean, where maps of the coastlines indicate that the area is prone to tsunamis; it has been sensed by NASA from outer space; a ship's captain has noticed a telltale rise in the ocean around his boat and radioed this information to his shipping company; and beach vacationers have noticed that the tide has receded dramatically, as seen in Figure 6.1. All of this information is stored somewhere and in itself does not have a lot of meaning to disaster relief personnel in the countries surrounding the Indian Ocean. However, as a whole, it may clearly indicate a natural disaster with subsequent con-textually driven responses.

In the above example, a user may know to go to a database and retrieve data about one type of information about tsunamis. However, he or she may be unaware that other types of data not in the database are available that may provide useful and potentially critical information. Even if users are aware that other data exist, they may not be able to locate them.

A goal, then, of context data management is to develop a way to manage all the types of contextual data and semantic-processing rules that can be found in the context model. This can be done by aggregation of context data objects and related contexts into *sets* of data based on similarity and relationship by theme. This approach, the *set management approach*, can logically associate all related data for a type of event and select the appro-priate contextual set based on user-supplied criteria.

The rest of this chapter describes the development of the set paradigm, a new type of modeling notation to manage sets and support for selection operators for retrieval of contextual sets where ambiguity in selection because of coverage overlap may exist.

6.3 CONTEXT-ORIENTED DATA SET MANAGEMENT

Current approaches to data management assume that each individual piece of data, or data file, must be managed. For example, in a relational database, a mountain object would be stored in one row of the table containing mountains. Attributes of the mountain might be stored in a different table. In object-oriented methods, a mountain and its attributes might be stored in a single object. This approach works well in a homogeneous environment where all the data being managed are owned by the application that is managing them. Specifically, the application knows the format of the data it owns and thus manages each and every piece directly. However, this is not practical or in most cases feasible in a heterogeneous environment that includes a multitude of different applications' data. Applications cannot generally read and interpret each other's data. Nevertheless, while data may be for different applications and in different formats, they can still apply to the same geospatial region or problem. When this occurs, there is a need for the creation of information about relationships among the data, but no mechanism to build the relation because of the differing formats.

The above type of problem can be solved by a shift in the approach to how data are logically thought about and organized. Instead of attempting to manage individual pieces of data (e.g., mountain and attributes of mountains), which incidentally may be very complicated or impossible in a heterogeneous data format environment, one can make the approach less specific and less granular. The key is to manage contextual data on the thematic attributes describing contexts, those of time, space, impact, and similarity. In this approach, specific data objects in a context feature vector are not managed directly, but instead they are organized into sets where the set is the fundamental level of context data management.

The shift to set management for contextual information produces benefits that address other problems with managing data. Specifically, sets can be copies of a base contextual set. These sets can thus be lineages or versions of the base set. Once versions of context sets are established, each set can become a particular view of the data included in a context. When views and versions become possible as a result of this approach, then so do multiple information-consuming entities with their own lineage trees and domains of control for the sets they define and own. Extending this concept, it is possible to see that multiple views serving multiple users do a very thorough job of addressing the previous users' data coupling, which is defined as users modifying their own data and often working on overlapping spatial or temporal themes. Thus the benefits of this approach can have a large impact on a variety of problems. Because the set paradigm is data format independent, this is a robust approach. The addition of new formats of data that can be described in a context as they are developed will not in theory cause the new approach to degrade as would potentially happen with current approaches. Instead, one simply adds the new-format data file to the set without any impact on the management and retrieval of such data.

Key to the contextual set management paradigm is the need for a way to define and document the set abstraction, and the application of metadata in the set paradigm can be helpful for this purpose. A robust approach requires a robust method of description that is not corrupted by the addition of new data formats or other changes. Metadata are a description of data; therefore, a logical entity referred to as a *set* would be modeled and have its characteristics described abstractly. This description can be captured and managed in a data model that employs metadata to define and manage the sets of contextual information under their control. In such a model, a context's description, manipulation, and retrieval are based on meta-discretional information about the set of data in a supercontext.

Finally, while a set-based model can be useful, it can introduce retrieval ambiguity problems, such as having multiple members in the set or among different supercontext sets that cover the same spatial, temporal regions or impact knowledge space. This can be referred to as *retrieval ambiguity* and can be mitigated through the application of fuzzy set theory on relations in the set-based metadata model that manages the set abstraction. Fuzzy set theory can be utilized to make a generalized comment about the degree of possible membership a particular data file might have in a set covering a specific physical or abstract space, thus suggesting to a user of such information which supercontextual data set might be most appropriate for a thematic query. This issue is examined in more detail in the following section.

6.4 CONTEXTUAL SET SPATIAL AMBIGUITY IN RETRIEVAL

Data set ambiguity in contexts refers to the fact that for a given query or selection, it may be impossible to select an exact or correct match for the query because multiple sets of context data may satisfy the query fully or partially. For example, if a query is interested in all data about an event found at a given spatial location, multiple sets of contextual data may have overlapping boundaries that the location can fall inside of, and thus it becomes difficult to determine which set satisfies a query in the best fashion. Another example can be found when one considers the spatial data in a context where the boundaries of objects are approximately but not precisely known. For example, if one is mapping the extent of the spreading wave of a tsunami, the edge of the wave and thus its boundary may be one meter wide or it may be considered to be hundreds of meters wide. The selection of information about the edge of the tsunami wave can then become an ambiguous problem. Because multiple contextual sets about a given tsunami's boundary may exist, perhaps one defines the edge to be one meter, and the other for the same geographic location defines the edge to be 100 meters; the question can become "Which context set's data should be retrieved for a query?" Because contexts have multiple dimensions that define them, this problem can also exist for the temporal dimension of contextual sets and the impact dimension. Interestingly, the similarity dimension might be utilized on relations to improve query accuracy.

Ambiguity in contextual data sets impacts their use in a fairly significant fashion and has been analyzed previously for spatial data but not for the new and additional dimensions of context set data. Contextual data sets (CDSs) could be organized into large databases. Users of the database could then create spatial or geographic queries to the database to retrieve CDS data that are of interest to them based on geographic regions, extent, and so on. This process of doing this is sometimes referred to as *geographic information retrieval* (GIR) if the queries are for spatial data. GIR seeks to deal with spatial uncertainty and approximation in the methods by which traditional spatial data are indexed and retrieved.

To provide some background on GIR and how a contextual model needs to address different issues, it is key to note that at the center of GIR research are spatial queries. Queries are the expressed needs of users that drive the search for known objects in a database as well as relevant unknown items. Key to this is an indexing method that supports efficient retrieval of spatial data. In determining what is to be returned for a query, the elements of the query are matched against the index using algorithms. The mechanism for doing this is sometimes referred to as a *retrieval model.* It is the role of the retrieval model to resolve ambiguities in queries and what has been indexed to attempt to satisfy a user's query. Methods for doing this include probabilistic and deterministic techniques.

Models are referred to as *deterministic* when they calculate probabilities of *goodness of fit* to the query and return the results. In contrast, a deterministic model requires the exact match of an indexed or georeferenced object to the query. Oftentimes, a deterministic model will also include Boolean logic operators such as AND and OR, as are often found in current database SQL query systems.

Between the retrieval model and query processing in a spatially oriented system is the GIR indexing scheme, sometimes referred to as the *georeferencing* of the objects. Indexing schemes are critical to resolving spatial ambiguity problems. They must be able to address spatial queries in an unambiguous fashion, such as

- Point-in-polygon queries,

- Region queries, and

- Distance queries.

There are many methods of indexing spatial objects; one of these is referred to as the *quadtree decomposition method*. This scheme seeks to parse a spatial region, referred to as a *supercell*, into a collection of subcells. This is done through a process referred to as the *hierarchical decomposition of space* [7]. The subcells can then be indexed and sorted for the objects they contain.

Indexing schemes cannot deal with ambiguity issues well because they are not granular enough (e.g., the subcells cover too large a space), and then what is returned for a given query may be not what was expected. The result of the query can be an ambiguous set of objects. Ideally, then, cells in an indexing scheme would be only slightly larger than each geographic object in the database, and all objects would be a uniform size. Subcells then would be decomposed to cover only one object, and thus ambiguity would be eliminated in GIR. However, this is not practical for a variety of reasons.

A key shift in the new model for CDS management is toward being *less granular* than the GIR approach in the management of CDS data. Instead of managing geographic entities such as one might find in a GIR database, or for that matter the dimensional entities of temporality and similarity, the smallest unit of management in this approach is a single covering logical device, that of the concept of a *set* for CDS data management. This shift to being less granular has a variety of benefits, but it can introduce further ambiguity, as mentioned earlier in selecting and defining sets with multiple overlapping coverages. A *coverage* is the data found in a CDS that describe the dimensions of space, time, similarity, and impact. With this in mind, ambiguity can be defined as a condition where multiple tracts of CDS data may satisfy a given query for a particular geographic location, point in time, type of similarity, or type of impact. When this is the case, there is a need for mechanisms that will suggest to a consumer of contextual information which is most likely to be the best set to utilize for a thematic query.

To illustrate the above point graphically, consider the case where two contextual data sets have a spatial dimension with overlapping coverage for the origin of a tsunami, as shown in Figure 6.2. One context data set comes from satellite imagery, and the other from sensed information sources. A query for information about the origin of the tsunami is referred to as a *point-in-polygon query* and will return CDS A and B with no suggestion about which might be more relevant to the theme a consumer might be interested in. The expected ambiguity between trying to select between these coverages can be seen in Figure 6.2.

In Figure 6.2, a question arises about whether one wants to use CDS B or CDS A for a query about tsunami CDS data at time T_0. This is an example of a problem that is found with standard storage models for spatial data and is present in the set-based method. Ultimately, the choice of which set to choose in an ambiguous problem should be left up to the consumer of the information. However, application of fuzzy set theory and

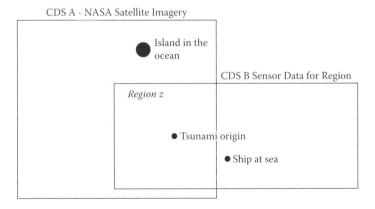

FIGURE 6.2   Example of data set ambiguities for spatial coverage of the origin of a tsunami.

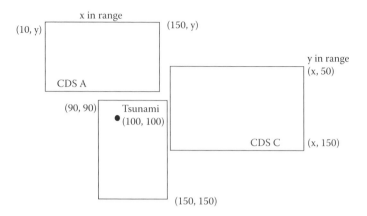

FIGURE 6.3   Initial range check to determine inclusion of data sets.

computational geometry can be applied to presenting potential data sets in ways that might mitigate the problem in Figure 6.2 to some extent.

Algorithmically, the first step to the solution for Figure 6.2 must be to identify the CDS data sets that potentially might solve the query for set data about the tsunami. This should start out by doing a simple range check to see if the location of the tsunami's y coordinates is within the y extent of any data set known to the system. An example of this is illustrated in Figure 6.3.

After some examination of the coordinate pairs, it is clear a simple range check on CDS A and CDS C eliminates them from inclusion as a candidate data set. This is because their x or y coordinate extents do not intersect those of the data set enclosing the tsunami point. After this first pass, the next step would be to determine inclusion of a point in the polygon. This can be done using a variety of mathematical formulas. The preferred method for its simplicity is to calculate the cross-product between a vector to the point being tested and a vector defined by a boundary, as shown in Figure 6.4.

The calculation of the vector cross-product is given by the following equation:

$$V1 \times V2$$

which can be expanded for a two-dimensional vector to

$$z = (V2x1 - V2x0)(V1x1 - V1y0) - (V1x1 - V1x0)(V2y1 - V2y0)$$

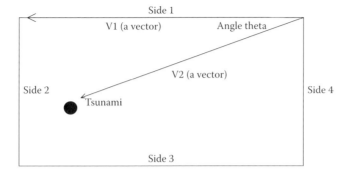

FIGURE 6.4   Calculation of a vector cross-product to determine point inclusion in the bounding polygon.

Because $V2 \times V1 = -(V1 \times V2)$, the value of z will be less than zero (< 0) when moving around the rectangle in a counterclockwise direction.

The next step for calculation of containment or range checking would be to simply calculate z for all sides while stepping around the bounding polygon in a counterclockwise direction. If the point representing the tsunami's origin is inside the bounding rectangular polygon, all values of z will be negative. If the tsunami is on an edge of the bounding polygon, the value of $z = 0$. Finally, if the tsunami's origin is outside of the polygon, one of the edge cross-product calculations will return a value greater than zero (> 0).

Using this technique, it is possible to establish that a spatial coverage for a CDS does include the spatial point in question. Without much modification, this technique can also be made to work for regions delimited by polygons that are entirely or partially contained by a data set's bounding polygon. Once coverage has been selected as a potential solution to a query, the next step could be to apply fuzzy set theory to rank the relevance of the coverage to the point of origin of the tsunami. This is the application applied later in the development of the storage model to manage contextual data.

Utilization of fuzzy sets can be done in one of several fashions. The first of these might be to calculate the distance between the tsunami's point of origin and the centroid of a bounding polygon. The distance value could become a component in the return value for the fuzzy membership function for the spatial aspects of CDS data, $C_{Spatial}()$, which is discussed later. In this case, the smaller the distance, the more centrally located the point representing the tsunami's point of origin (the theme) is to the coverage

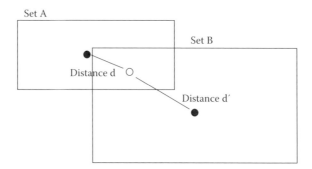

FIGURE 6.5   Application of distance as a return value for a fuzzy function.

being considered, which thus may make it more fit in this simple example to be considered the best solution to the query. This scenario is shown graphically in Figure 6.5.

The value of the fuzzy membership function $C_{Spatial}()$ to select for Set B may be .9, whereas the value of the fuzzy function $C_{Spatial}()$ for Set A might be .3. Higher values of $C_{Spatial}()$ would reflect the fact that the distance to the nearest centroid was smaller. This would then suggest that because the point represented by the tsunami's origin is more centrally located in Set A, this contextual set is a much better set to use.

While this approach makes sense, it is a bit simplistic. Therefore, weighting factors in conjunction with attributes of the data sets themselves might also be considered. An example of this might be that both sets have metadata for *data accuracy,* and the *time of last data collection.* In a CDS set model, one might argue that the data accuracy is very important (.9) because improved accuracy would be expected to reflect current reality better. Following this logic, a scheme then might weight *time of last data collection* very heavily. Less weight might be given to accuracy, especially as an adaptive function of the time of last collection (.3).

A weighting scheme to utilize in retrieval might develop in the CDS set model as a function to assist in resolving data ambiguity, and in this case might become

$$C_{weight}() = distance^* (\ .9^*days\ since\ last\ edit + .3^*time\ of\ day\ edited)$$

In the above scheme, geometric properties have been combined with metadata attribute properties for a given CDS set to solve ambiguous retrieval problems.

6.5 A SET MODEL-BASED ERD

In previous work on the subject of temporal and spatial data storage and retrieval ambiguity [11], an initial data model was developed that manages sets of data utilizing some of the concepts previously mentioned. This model is based on the aggregation of data into sets of related information and then management of the sets, their storage, and retrieval utilizing metadata. The model is shown in Figure 6.6.

To extend the capabilities of the model to manage a variety of data formats, this model is meant to be populated with attributes utilizing metadata that describe the organization and management of the CDS sets. The benefit of using metadata is that it is abstract and therefore resistant to needing continual redevelopment of the model as data being managed in the sets changes in format.

In Figure 6.6, the entities were defined to be the following:

- The *SuperSet* entity defines collections of contextual *Subsets*. It has a recursive relationship to other *SuperSet* instances. The relationship *related* shows the fact that multiple supersets may be related to each other. For example, supersets may cover the same dimensions of multiple contexts.

- A *SuperSet* is composed of multiple *Subsets*. Using the example of the tsunami, a superset may be composed of all the CDS sets of tsunami

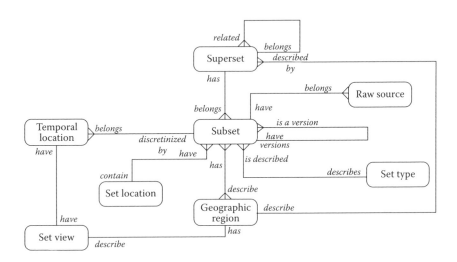

FIGURE 6.6   Initial ERD model for set management.

data that have been created over a given period of time. The relationship of supersets to subsets introduces the new extension of fuzzy subsets. An instance of a subset entity is a single logical metadescription of a component in a superset. It also has a recursive relationship to itself that allows the user to implement versions of the subsets and thus versions of the sets, that is, a lineage tree.

- The *Temporal Location* entity represents a locational time definition for a data set. It has temporal attribute values and geospatial coordinates that collectively create identifiers for a particular data set.

- The *Geographic Region* entity locates a *Subset* of data by the type of spatial coverage it has. This entity works in conjunction with the *Temporal Location* entity to locate a set of data in time and space. The rationale is that for a given spatial area, there may be multiple coverages generated over time. Therefore, the problem becomes one of locating spatial data in space and time.

- The *Set Type* entity describes the type of data a *Subset* may contain. An example of the expected types were *image*, *raster*, or the like. This entity was also a candidate for fuzzy notation extension.

- The *Set View* entity provides a repository for information about various views of data that may exist for a given *Subset*. *Set View* makes it possible to have multiple views of the same data set. Such views would differ by such things as a data set's perspective, scale, or projection.

- The *Location* entity describes the physical location of the *Subset*. This entity is required because of the need for a model where data can be distributed around a computer network. This entity provides the potential to support distributed database mechanisms because parts of the database are not in the same physical data space at all times. It also provides a way to have alternative views of data that are not centrally located.

- The *Raw Source* entity is the only entity in this model whose instantiation represents a real, nonmetadata object, in this case the physical data described by *Subset*. This entity contains enough information to locate and identify a file as belonging to a particular instance of a subset.

6.6 A FUZZY ERD MODEL FOR CONTEXTUAL DATA MANAGEMENT

The data model in Figure 6.6 provides an initial foundation to address problems inherent in the management of context data. The data model was developed to describe contextual information using metadata. One of the problems with contexts is ambiguity in the selection of data. These problems can be addressed by the application of fuzzy set theory. For example, several overlapping coverages for tsunami information could exist. In this sense, overlapping coverage is defined to be multiple contexts with information fully or partially about the same event, for example the origin of the tsunami. Keep in mind that there is a tendency to think of such information as geospatial, but it may also include images and textual descriptions. The key concept in retrieval is to find the most appropriate coverage for a given query. *Appropriate* is a term that can only be defined for the consumer of the information, the user. The logical question becomes which context data set to select and use for a given purpose.

Overlapping contextual spatial coverages are a type of ambiguity. We have also identified that there can be overlapping contextual temporal locations. We can also have overlap in the similarity of contexts and their impact dimension. When considering the problem of overlap in the selection and retrieval of information, the types of overlap that can be present must also be considered. Overlap can be partial or complete overlap, with different descriptive characteristics to coverage such as different projections, scales, and data types. These can also become complications to ambiguity of selection. Considering this situation, it is clear that ambiguity on an attribute of spatial data can compound with other ambiguities about the same data. This can have the potential of leading to much larger ambiguities.

To date, much work has been done in the development of fuzzy set theory and techniques for decision making such as using open-weighted operators (OWAs) [10]. Little of this work has been applied to the management of contextual data. In particular, theoretical discussions need some form of implementation to solve real-world problems. What is needed is the application of theory and a representational notation. The application could then be used to solve real-world problems such as data ambiguities. The Figure 6.6 data model can be extended with new types of fuzzy operators that address ambiguous selection problems to create a more powerful model that can deal with the problem of ambiguous data.

Chen [3] defines an extended entity relation (EER) model to consist of the triple

$$M = (E, R, A)$$

where M represents model, E represents entities, R represents relationships, and A represents relationships. E, R, and A are defined to have fuzzy membership functions. In particular,

$R = \{U_r(R)/R \;|\}$ where R is a relationship involving entities in Domain(E) and $Ur(R) \in [0,1]$

In this case, $Ur()$ is a fuzzy membership function on the relationship between two entities in a data model. Chen defines fuzzy membership functions on attributes and entities as well. Because of the above, it is possible to have fuzzy relations on relations, without built-in dependencies on other types of fuzzy objects in a model. Based on this work, the proposed model extends the Figure 6.6 data model ERD to defining notations that describe the application of fuzzy theory to relations.

In the next section, we will introduce new notations and provide a rationale for their existence within the context of the ambiguous selection problem domain. Finally, we will apply the new notations to the Figure 6.6 ERD for set management, thus deriving a new type of model that merges standard ERD notation, metadata management of data with fuzzy theory to organize and select contextual data, and descriptions of sets of contextual information. The model also provides architectural locations for the components and dimensions of contexts that have been previously introduced.

6.7 CONTEXTUAL SUBSETS

The first new notational convention is the context subset symbol. The subset symbol defines a new type of relationship on an entity that borrows from object-oriented constructs: the *bag*. An entity with the subset symbol defined on one of its relations is a nonunique entity, unlike most entities in an ERD model. The rationale for its existence is that multiple copies of a subset containing the same elements can exist for different overlapping temporal, spatial, impact, and similarity coverages for a given event (e.g., the tsunami). This circumstance can occur as a result of various versions of the same *Subset*, or normal editing operations. The symbol is defined graphically as

$$\subseteq$$

By the nature of being a nonunique entity, a relationship with the subset definition also becomes a fuzzy relationship. This is due to the fact that when one desires to view a subset, the question becomes which one should be selected.

6.8 FUZZY RELATION SIMILAR *FNS()*

Fuzzy theory literature [2] defines a membership function that operates on discrete objects. This function is defined as similar and has the following property:

$$FnS() = \begin{cases} 1 & \text{if } a \in \text{domain}(A) \\ 0 & \text{if } a \notin \text{domain}(A) \\ [0,1] & \text{if } a \text{ is a partial member of domain}(A) \end{cases}$$

This function expresses a dimension of contextual processing, that of similarity, and thus provides a framework to reason about impact. The function is also useful in contexts and supercontexts, where overlapping coverages of the same event space or time may exist, but some coverage for a variety of reasons may be more relevant to a particular concept such as a desire to perform editing of surrounding regions. Similarity can also be nonspatiotemporal, and thus this function on a relationship can also model thematic similarity of what might appear to be disjointed supercontexts. The actual definition of how partial membership is calculated and how spatial models are defined has been the subject of much research, including the application of OWAs [1,2,4,5,8,9] and the calculation of relevance to a concept [6].

The data model being developed for a contexts model in this chapter initially seeks to provide alternative view support for overlapping geospatial coverages. Because of the ambiguities induced by this, we introduce a notation that represents the fuzzy relation resolved by the definition of the function *FnS()*. This symbol is referred to as the fuzzy relation *FnS()* symbol and may be displayed in an ERD model along with the relations between entities. *FnS()* function is evaluated when a query on a relationship with the *FnS*() function is ERD generated. The query returns a ranked set of items with a similarity value in the range of [0,1].

Another property, albeit an unexplored one, of the function *FnS*() is that it reflects the fuzzy degree of relation that entities have with other entities. Therefore, it can also be a metaphor for the degree of confidence a modeler might have in knowledge that similarity is actually known. Low or no confidence (e.g., $tr = 0$) should drive the membership to zero (non-membership). In a model where an entity instance may be selected via various related entities, the effect is that the entity with the strongest relationship to another given entity will produce the highest similarity values and thus provide a better selection.

6.9 FUZZY DIRECTIONALITY

Fuzziness in the context data model is not bidirectional on a given relationship. In other words, when executing a query, the function should logically only be invoked in the selection of instances in a target entity. It may not make sense to invoke the function in a reverse direction of query. Therefore, there needs to be some indication of the direction that a function is applied when processing a query. This is denoted by the inclusion of the following arrow symbols on the fuzzy relationship. These arrows are found to the left of the fuzzy symbol and point in the direction that the fuzzy function *FnS*() or $C_{spatial}$() applies. If a fuzzy relationship is defined in both directions, which implies a type of *m:n* relation, the symbol is a double-headed arrow.

Directional fuzziness for the *FnS*() or $C_{spatial}$() function points in the direction that the function is applied for selection and is denoted by the following symbols:

$$\leftarrow, \uparrow, \rightarrow, \downarrow$$

Bidirectional application is a member of the class of *m:n* relations and is denoted by

6.10 DISCRETIZING FUNCTION $C_{TEMPORAL}$ ()

Because of the data ambiguities mentioned previously, the new context-based data model uses time as an attribute in describing data. However, this leads to temporal ambiguities in the selection of a *Subset* of data because the *Subset* can exist at many points in time. However, there are certain points in time where the relevance of data to a concept or operation (e.g., a selection) query is more relevant. Therefore, the relation of *Subset* to *Temporal Location* in the initial version of a data model can have fuzzy logic applied. Time is not a discrete value; it is continuous data in nature. Some attempts have been made to discretize continuous temporal data by Shekar in the Spatial Data ER model [9] using the *discretized by* relation. But no known attempts have been made to deal with this in a fuzzy fashion This leads to the need to define a new function $C_{temporal}()$ that could be used to calculate discrete fuzzy membership values over continuous fields, including time.

If time is nondiscrete, then a function needs to be developed that converts continuous data into discrete values so that it can be utilized in queries which use discrete values. The question becomes how to represent continuous data in a fashion so that a function can make computations on the data and return a discrete value representing membership. Upon examination of this issue, nondiscrete data can be defined as a bounded range $[m,n]$, where the beginning of the continuous temporal data starts at time m and terminates at time n. This representation then makes it possible to develop the function $C_{temporal}()$ and its behavior over continuous data.

In this function, one may want to think of a window of time in which a set of context data was created at time t_m, spanning to a point in time where the data are no longer modified, t_n. The range $[t_m,t_n]$ is referred to as the *window* because it can slide backward and forward, growing and shrinking over continuous field data that one seeks to determine the degree of membership of a selection point of time t. A function that can then be used to retrieve relevant sets of contexts can be defined as

$$\text{INRANGE}([t_m,t_n], t) = \{ABS\,[(t - t_m)/(t_n - t_m)]\}$$

where $ABS()$ is simply the absolute value of the calculation.

The effect of $C_{spatial}()$ is to make a discrete statement about nondiscrete data, which makes it possible to make assertions about fuzziness and possibilities. The statement is of course relative to the bounded range $[m,n]$, and therefore $C_{spatial}()$ should be formally denoted for modeling notation as

$$C_{spatial}()_{[m,n]}$$

6.11 FUZZY RELATION $C_{SPATIAL}()$

The function $FnS()$ is not a complete function for the solution to selection problems in our ERD model. This is because it does not consider the centrality of a point P composed of an x, y, and perhaps z component that one wishes to retrieve context data about. This led to the creation of a function referred to as $C_{Spatial}()$. The characteristic function $C_{Spatial}()$ needs to contain a function that measures distance, a new term d that can be derived in the following manner:

$$d = \min\left(\sqrt{(centroidx - point\ x)^2 + (centroidy - point\ y)^2}\right)$$

$$centroidx = .5 * (s1x2 - s1x1)$$

$$centroidy = .5 * (s2y2 - s2y1)$$

The above equation refers to a rectangular bounding hull created around a geographic coverage. $Centroid_n$ is the centroid of the bounding hull found by finding the midpoint of side 1 for $centroidx$ and the midpoint of side 2 in y for $centroidy$.

$Pointx$ and $Pointy$ are the coordinates of a spatial entity or center of a region of interest that one is seeking the most centrally located coverage for. The d value in the a characteristic function for $C_{spatial}()$ then becomes a measure of the minimum distance of a coverage's centroid to a spatial entity. The effect of the equation is to find a context's coverage that is most central to the spatial entity of interest. The goal is to weight the characteristic function values with a measured degree of centralization to a spatial center of interest when selecting fuzzy data for a particular problem.

6.12 EXTENDED DATA MODEL FOR THE STORAGE OF CONTEXT DATA SETS

With an understanding of the issues found in the retrieval of context data sets and some proposed characteristic functions, a new data model can be presented for the management of sets of context data. This model considers the vagaries of ambiguity in time and space selection as well as the dimensions of similarity and impact. The model also needed to be extended to include the dimensions of contexts. These include *impact* implemented as an entity and *similarity* as a fuzzy set function applied on a relationship. Many notable modifications of the previous data model are apparent, as shown in Figure 6.7.

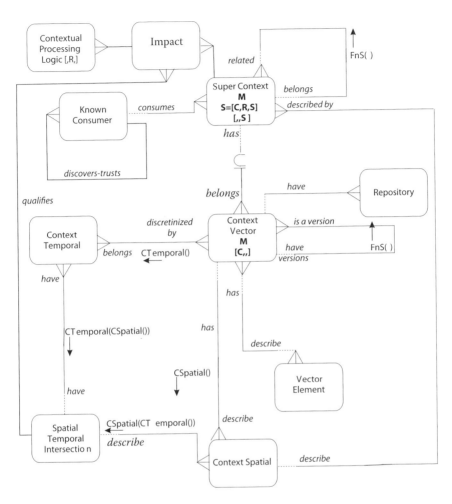

FIGURE 6.7   Extended-context set management model.

The first addition to the original model of note is the notation of M being added. This notation could have been place in many entities but is particularly relevant to the context vector entity. M means that the entity *primarily* contains metadata, in this case data about a given context vector. The logic is that the *Vector Element* and *Repository* entities contain the actual raw data, so therefore the context vector only contains descriptions of these data. Because *Super Context* is a collection of *Context Vectors*, the logic follows that it is also a metadata entity. Of note, in the listed super-context tuple $[C, R, S]$, this entity also becomes the architectural location for security information, which is discussed in Chapter 7.

The relationship between *SuperSet* and *Set* in the original model had a subset notation, and the name of the entities have been changed in the new contextual model. Additionally, a weighting operator (OWA) represented by a summation sign has been removed because a supercontext only consists of all the context vectors that aggregate to define the supercontext. The SuperSet entity has now become *Super Context* and the original *Set* entity is now *Context Vector*. *Context Vectors* by definition are not unique entities. When considered with the entities with which they have relations, however, they can become unique.

A subset of a *Super Context* data has an unusual property in that this entity is not unique in itself. It becomes unique when the fuzzy temporal, spatial, and impact relations around it are considered. There can be multiple physical files containing context data and rules that a subset may represent.

This relation can be characterized by the existence of multiple *Vector Elements* of heterogeneous data formats that cover the same area but may be of different scale or perspective. Each one of the *Context Vectors* may cover or be relevant to a minute part of the spatial coverage of a *Super Context* and is therefore a subset. Additionally, the *Context Vector* coverages may not be crisply defined in the spatial sense. They may also cover other areas defined as part of other partitions in the superset. This leads to the property of sometimes being unique and sometimes not being unique.

The *Context Spatial* entity has a $C_{\text{spatial}}()$ relationship with *Context Vector*. The rationale is that because *Context Vectors* of data can be overlapping in spatial coverage for a given point in space, selection of a subset becomes an ambiguous problem. The $C_{\text{spatial}}()$ symbol then implies that selection of *Context Vectors* covering a geospatial point needs to be done using some type of fuzzy selection. Of note, the directional indicator means that given *Context Vectors* data, the $C_{\text{spatial}}()$ function is executed on *Context Spatial* to find all the spatial context data that may apply to a *Context Vector*.

The $C_{\text{temporal}}()$ relation between the *Context Temporal* relation and *Context Vectors* can be used to select *Context Vectors* that cover a given range in time and thus is used to select the most relevant *Context Vector*. This function exists due to the presence of long editing transactions on the data. A *Context Vector* is not instantly updated during a long transaction. Certain points in the existence of the *Context Vectors* may be more of interest when selecting a set to view, but all are valid descriptions of the subset. The $C_{\text{temporal}}()$ function exists as a sliding window of possibility for selecting a *Context Vector* that has existed and was updated over a period of time. This allows one to select the *Context Vectors* in a specific time range.

The *Temporal Region* entity exists because there is a need for given data sets to map to various locations in time and spatial coverages. The *discretized relationship* was originally defined to map a value to a continuous field. In this case, the discretized relation has been extended to represent a discretized function where the continuous field is time. Because this function can relate a data set to various points in time and coverage of several different spaces, this function is a fuzzy function, $C_{temporal}()$, that selects time and spatial definitions for a *Context Vector*.

New in the model is that the dimension of *Similarity* is applied as a directed fuzzy function on relationships. This allows *Super Contexts* and *Context Vectors* to be associated. Of particular use is that *Super Contexts* can be similar to other instances; thus, multiple types of thematic objects (mentioned previously) can be supported in the model.

Vector Element is a new entity in the model. Because a *Context Vector* can be composed of infinite numbers of elements, there is a need to describe each of the elements and the potential processing of idiosyncrasies. *Vector Element* is where this information would be placed.

Repository is a location where the physical storage structure might be described. Such a structure maps the abstraction of set management onto a physical system. Systems that might be described here are the type of database (ORACLE, DB2) or file or file locations.

The entity *Spatial Temporal* location is an intersection entity that is meant to capture the notion that contextual data can only be uniquely identified by their intrinsic descriptors of the time and the location, or space, that the thematic object described by a *Super Context* existed in. Such a tuple can be assigned something like a globally unique identifier (GUID) in this entity, but it is the temporal and spatial identifiers that localize the *Super Contexts* theme.

Related to the above argument is that time and space will have consequent impacts, a dimension of contextual data, associated with them and the *Super Contexts* theme. For instance, the notion *calm Indian Ocean one week ago* has a very different potential impact associated with it than *Indian Ocean with spreading tsunami one minute ago*. Thus, the impact entity has been related to these concepts and entities in the model.

Known Consumer is the place where hyperdistribution information for contextual data would be stored. On the recursive relation, the concepts described in the chapter about the discovery and trust of consumers in their process of getting to know each other for hyperdistribution are supported.

Finally, the end product of most of the discussion about contextual processing is the processing part. With the dimensions of contextual processing supported in the model, and tied together through relations, the *Contextual Processing Logic* entity is where the semantic rules for a given impact would be contained. This, like the *Repository*, is where there is a physical interface to the outside world where users define the specific actions for their systems for a given set of contextual reasoning and semantic meaning.

6.13 EXAMPLE: SET-BASED MODELING AND CONTEXTUAL DATA MANAGEMENT

Currently in IT architectures, data are stored in relational databases which have a hard time storing and managing the relationships among information, especially if they are complex such as those found in temporal and spatial data. Considering the other dimensions of context, similarity, and impact, current technology tends to run up against its dated limitations. For instance,

- It is difficult to know that one type of spatial data is *related to another*.

- It is difficult to determine which type of spatial or temporal data to retrieve if they *overlap*.

- It can be difficult to find *all* the data about a given theme, especially if they are hyperdistributed across multiple computers.

- Modeling of *similarity* between data might be implemented but is not natively supported.

- Keeping *relationships among concepts* of the impact dimensions is undetermined.

The *set-based data model* proposed for contextual processing can easily address all of the above because instead of micromanagement of data such as that found in relational technology, it manages a much coarser granularity, that of the set. Set-modeling concepts can be implemented on top of existing relational technology or for that matter any type of existing management technology.

In this paradigm, a user may be interested in all the data related to a theme such as tsunamis based on the selection criteria of temporality, spatiality, impact, and similarity. The proposed model has the capability to search across a hyperdistributed universe of information and return sets of information based on the degree that they are related to other sets about the same theme. The application of fuzziness and some of the other concepts in this proposed model could be further developed such that the returned groups of sets might be ranked by degree of applicability to the original request for information. The underlying set model maintains the internal relationships that support the retrieval by a *degree of relationship* process. This capability is currently not found or is seldom found in current implementations of information repositories.

6.14 RESEARCH DIRECTIONS IN CONTEXTUALLY BASED SET MODEL DATA MANAGEMENT

The ideas and concepts introduced in this chapter offer one potential method to manage contextual information. They should be examined and enriched with more rigorous definitions. Additionally, there is potential to define metrics that might express such things as the following:

- Confidence in relationships

- Trust in relationships

- Model quality metrics

The model proposed in this chapter is architectural in nature and thus not attributed. The process of determining the metadata to attribute the model could be an active and vibrant research area pursued with the question of how correct attribution will be determined. One potential approach to this question is to use an established standard such as the geographic and military communities currently have as a starting place.

Development of the retrieval mechanism might be another strong area for research. In this vein, definition and refinement of the fuzzy selection and relation mechanisms could be pursued. There will be a need for development of the mathematics behind the concepts. Such equations probably should have an adaptive mechanism, making it possible for them to change behavior based on the context of environmental data. This could lead to the research areas of *contextual set management* and *contextual retrieval of information*.

Finally, it would be interesting to study the development of methods that could allow the model to dynamically redefine relations and data based on the context in which information is being managed. Such a model might be based on biologically inspired algorithms or complexity theory and could be very appropriate in view of the subject matter being managed.

The next and final chapter of this book will address the very complex issues of securing a voluminous, streaming, and infinite quantity of data as they are generated and move from producers of contextual data to consumers. The argument is made that perhaps not all data need to be secured and that a method for cutting down on securities overhead might be to model security based on *relationships among data*. A new model is presented that utilizes the complex mathematical properties of branes from cosmology to suggest the level of security that contextual data might need and allow the consumers to determine how to implement that security. The paradigm of contextual security is developed.

REFERENCES

1. Barker, R. (1990). *CASE*Method: Entity relationship modeling*. Reading, MA: Addison Wesley Longman.
2. Burrough, P. (1996). *Natural objects with indeterminate boundaries: Geographic objects with indeterminate boundaries*. Philadelphia: Taylor & Francis.
3. Chen, G., and E. Kerre. (1998). *Extending ER/EER concepts towards fuzzy conceptual data modeling*. Beijing: MIS Division, School of Economics & Management, Tsinghua University.
4. Cross, V. (1996). Towards a unifying framework for a fuzzy object model. In *Proceedings: FUZZ-IEEE*, pp. 85–92. Piscataway, NJ: IEEE.
5. Dubois, D., and H. Prade. (1985). *Theorie des Possibilities: Applications a la representation des connaissances en informatique*. Paris: Masson.
6. Morris, A., F. Petry, and M. Cobb. (1998). Incorporating spatial data into the fuzzy object oriented data model. In *Proceedings of the Seventh International Conference on Information Processing and Management of Uncertainty in Knowledge Systems (IPMU'98)*, pp. 604–611. New York: IPMU.
7. Osborne, H., and R. Applegren. (1998). *University of Idaho experimental forest* (information pamphlet). Moscow: University of Idaho.
8. Peuquet, D., and L. Qian. (1996). An integrated database design for temporal spatial data. In *Advances in Spatial Data Research: Proceedings of the 7th International Symposium on Spatial Data Handling*. Delft: Delft University of Technology.
9. Shekar, S., M. Coyle, B. Goyal, D. Liu, and S. Sarkar. Data models in geographic information systems. *Communications of the ACM* 40 (4): 103–111.
10. Yager, R., and J. Kacprzyk. (1997). *The weighted averaging operators: Theory and applications*. Boston: Kluwer Academic.
11. Vert, G., Stock, M., Morris, A. Extending ERD modeling notation to fuzzy management of GIS datasets. *Data and Knowledge Engineering*, Vol. 40, pp. 163–169, 2002.

CHAPTER 7

Security Modeling Using Contextual Data Cosmology and Brane Surfaces

THEME

This last chapter on contextual-based processing looks at a critical issue in global contextual processing: how the information in a context is secured. To understand the proposed model, we discuss the nature of the methods currently used in security and the nature of security on the Internet. The Internet, being essentially a flat peer-based model, has little security, yet contexts that are of a global nature either physically or logically and hyperdistributed need some methods for determination of confidence in the accuracy of contextual information before it is consumed. A new geometrically based mathematical method for reasoning about levels of security requirements is proposed through the application and development of a *brane*. Branes can be utilized to integrate the four dimensions of contexts—time, space, similarity, and impact—into a model for determination of security requirements required for a context. This is unified with the concept of *spot security* as a way to reason about the level of security that might be required for a context and the relationship of security to that of confidence in utilization of the context's information. This model is intuitive in application and provides a metric upon which current security methods can be applied without throwing out existing techniques or reinvention of current methods utilized in security.

7.1 GENERAL SECURITY

7.1.1 Cybersecurity Overview and Issues

The area of information security covers a wide range of topics related to computers and the information that is stored on them. This field has a rich and diverse set of subtopics ranging from policy development to authentication to detection of attacks on systems. The objectives of computer security vary but often are concerned with protection of the computational resources that process information and maintenance of the integrity of such information. Often the means and methods of implementation of security start from development of a policy document that can be ambiguous in its interpretation in regard to specific implementation. Because of this, there can never be common implementations that fit across all systems for all instances. This means that it is hard to evaluate whether a security implementation is correct or not correct. What can be evaluated is the application of established methods based on security-level requirements derived from policy.

Computer security policy and implementation often take the form of proscribing constraints on the operation, utilization, and application of computational resources. This makes development of good security difficult because a large number of security challenges often deal with anticipation of events that have not been encountered before and appropriate responses to such unknown events. Responding to the past is easy; anticipating the future is the challenge of security.

Complicating this fact is that requirements on different systems will vary depending on the system. Therefore, there is not and cannot be a one-size-fits-all approach. Yet, security has typically been looked at as a black-and-white paradigm. For instance, if a system is attacked, and the authentication system prevents unauthorized access, this is good: it is a white situation. If the same scenario occurs and the system allows unauthorized access, then the security system failed: a black situation. Even if a comprehensively effective and predictive policy can be developed, computers and their software are inherently different from most other types of systems because of their complexity. Complexity makes complete evaluation of the operation of computer hardware and its software under all possible circumstances literally impossible. The field of software engineering is interested in measurement of the correctness of the behavior of software, in particular to try to deal with the complexity of program design and operation, but has a way to go before it can be completely applied to security problems in a way that solves the intricacies of development and implementation of good security.

Because of the complexities found in systems, research has often turned to mathematical models and theories as the basis for development of security. These methods can range from Bayesian to statistical to probabilistic. Numbers that come out of equations are often considered to be clear-cut in their meaning. This is a paradigm where the mathematical results of trying to model security are referred to as *crisp*. For instance, the number 1 can be interpreted as the presence of an entity or an attack, and the number 0 can be interpreted as the lack of presence of an entity or no attack. As discussed in Chapter 2, this is referred to as *discrete data*. However, discrete data does not model the real world, which is often continuous in character and the types of information it generates and can be ambiguous in meaning. Yet, discrete reasoning is all over security thinking, right at the core practices and methods employed in security. In the *authentication* of information, an entity is what it claims to be or it is not. In *access control*, if a user knows a password, then he or she can log in (1) or cannot log in (0). When *detecting intrusions*, if a system has the number of packets increase in an N hour period on a port and a series of other events in a vector of binary variables, then the conclusion is that a system has been intruded. Some other applications of binary thinking in security are the following:

- Physically control access to computing resources via secured locks and facilities.

- Develop mechanisms in the hardware that control the operation of software and allow it to only perform safe operations.

- Create operating system mechanisms that impose rules on programs to avoid trusting computer programs.

- Develop methods of programming and code analysis that remove vulnerabilities in software.

Measurement of the level of security for systems has been developed such that systems can be evaluated relatively against each other to decide what level of security assurance, and thus security, exists on the systems. Examples of these are the common criteria levels EAL4 or EAL5. In such schemes, often certain levels indicate that the security functions are implemented flawlessly or are less dependable. These schemes are typically oriented toward implementation on servers and found in a wide variety of critical web resources.

To increase the security of a computer dramatically, the best practice is to not have it connected to the Internet. This is a reason why IBM computers show up less in reports as being hacked into as they are often facilities based and thus secured. Of course, this does not mitigate the insider threat problem but does cut down on some other common vulnerabilities that PCs share.

Most of this mind-set for how security is approached and methods of thinking about security is deeply rooted in the foundations of computer science from almost sixty years ago, a time when there was no such thing as an Internet. At that time, the concept of binary math to represent a voltage or lack of voltage in a circuit became the basis for most if not all architectural viewpoints on the construction of computers and the software running on them. So, it is not unusual to find that the approaches to security methods today are often based on the viewpoints of the past. The interesting point to consider is that if this viewpoint is still as valid as it was sixty years ago, why are the security problems on the Internet exploding? For instance, recently it was estimated that 80% of the network traffic and spam on the Internet was being generated by botnets, which may infect as many as 3 million computers at present and are growing out of control.

Botnets and most other types of threats currently found in the security pantheon are directly due to the hyperconnectivity of computational resources as a result of LANs, WANs, and the Internet. It is interesting to speculate what security challenges would be if the term and implementation of *connectivity* did not exist. Even more interesting, and another legacy of early computer science thinking, is the hierarchy. In a hierarchy, every entity is a parent of another entity. This largely originated from the concept of powerful machines such as mainframes being accessed by the minions of slower, less powerful machines for information. This also became another key concept in how security is often thought about. In most architectural thinking, a central authority, a computer, or a classified team of experts administers security. This is a bad architecture, as any person who has been in the military will generally tell you, because a direct hit on the central resource becomes a knockout blow for the entire system and is one of the reasons why DOS attacks are so effective against servers.

For instance, if one knows the location of the server, it is far more effective to concentrate on its destruction than to go after all the smaller machines communicating with the server because if the server cannot function, neither can its clients.

Interestingly, the Internet suffers from the concept of hierarchy in security thinking. There is essentially no security on the Internet itself because it is a flat model where information flows to wherever it is needed. At the point of consumption of information, one may find some type of hierarchy against which a security model has been overlaid, but the Internet itself does not have such a logical organization due to how it was architected. Nor can such an architecture be easily implemented due to political, legal, and technical issues.

All of the above complicate how security on contextual data might be implemented. With a context representing any type of logical or physical event, which can be streaming information continuously into a peer-based Internet environment where security does not exist and encountering hierarchal-based contextual information consumption nodes where security reasoning is thought of as 1's and 0's, the question of how to define a contextual security model becomes complicated and unique. The next sections will examine previous models and then present architectural and mathematical methods for approaching the new problem of context-based information systems security.

7.1.2 Models of Security

To understand what security might look like in a global contextual architecture, it is useful to review a few types of security models and the paradigms that currently exist.

A computer security model is a method for the specification and enforcement of the security policy of an organization or entity scheme. Security models are generally founded on the idea of access right, models of how the computer should operate, and the concepts of distribution as found in a network of computers.

There are several different types of models that have been created over the years. The access control list (ACL) is one of the most basic of models and has been around for years. An ACL is a list of permissions that a user may have on some entity in the computer system. Such a list can specify, for instance, that Bob can read a particular file.

The capability security model is created by having the ability to create a reference to an object which by virtue of possession gives the user a process to utilize the object in some way [1, 2]. An example might be found in opening a file given its name and the access rights on the file. If the opening is successful, an entry is made in a system table, thus granting the program that opened the file the capability to read or manipulate the file. Because the handle to the file is granted by the kernel, this is a type of security by the theory that other processes cannot directly access the file handle.

Multilevel security (MLS) allows a computer system to process and share information based on different levels of security [3]. In this type of model, users must have a clearance and a need to know in order to access information which the architecture of the MLS system enforces. MLS allows users with elevated security levels to have access to less sensitive information and allows users with enhanced-security-level clearance to share documents with less cleared users after sensitive material has been removed. In the MLS system, users with a lower clearance can share their information with users holding higher security clearances, but not vice versa.

Role-based access control (RBAC) is similar in concept to an access control list but it has the extension that access is based on the role a user plays in regard to information resources. Roles have certain types of operations that they are allowed to perform on data objects, and thus a user assigned to a role inherits the capability of doing the operations of the assigned role. It is a newer approach than mandatory access control and discretionary access control. RBAC can simulate discretionary access control and mandatory access control [6, 7]. MAC can also simulate RBAC if the internal data structures are restricted to being a tree [8].

Another type of security model was create by Biba in 1977 [9] and is referred to as the *Biba model*. This model is based on state transitions and security policies to control access to objects. In this model, entities consisting of information and people are organized into a system of levels where each level represents a class of integrity. The concept of the levels is such that data at a given level can never be corrupted from a lower level. This is accomplished by limiting the ability to write data to a higher level and not reading data from a lower level. This does not limit a user at a higher level from creating new data at a lower level. Roughly, the levels correspond to the integrity of information found at each level.

Finally, the object capability model is based on the concepts of objects that are referred to as *actors*. The ability to perform an action in this model is based on the idea that there is an address that is unforgeable and that it can be sent a message specifying an operation to be performed that is also unforgeable. In this model, there is the concept of connectivity in that the actors can only get new addresses to other existing actors via messages. If two actors want access to each other, there must be a chain of connectivity in existence. The advantage to such a system is that it inherently derives the benefits of object orientation by such things as encapsulation, information hiding, and modularity, which can be lent to applications in the security realm.

All of the above models can have application to the question of how a computer system is secured and how access control is done for a given site. However, they all are very specific and sometimes conflicting. With the nature of contextual information, there are the following properties that somehow must be included in a model or approach to contextual security:

- Spatiality

- Temporality

- Aggregation and ambiguity

- Limitations on computational power

- Hyperdistribution

- Political forces

- Impact and criticality

- Versioning

7.2 CHALLENGES AND ISSUES IN THE DEVELOPMENT OF CONTEXTUAL SECURITY

7.2.1 Elements of Contexts

To better understand the issues connected with security models for contexts, we review some earlier details about the newly developing model for contextual processing.

Contextual processing is based on the idea that information can be collected about natural or abstract events and that information surrounding that information, metainformation, can then be used to control how the information is processed and disseminated. In its simplest form, a context is composed of a feature vector

$$F_n<a_1,a_n>$$

where the attributes of the vector can be of any data type that can be collected about an event. This means that it can be composed of images, audio, alphanumeric data, or the like. Feature vectors can be aggregated via similarity analysis methods into supercontexts S_c. The methods that might be applied for similarity reasoning can be statistical, probabilistic (e.g., Bayesian), possibilistic (e.g., fuzzy sets), or based on machine learning and data mining (e.g., decision trees). Aggregation into supersets is to mitigate the collection of missing or imperfect information and to minimize computational overhead when processing contexts.

A complete context, or a *supercontext*, is described as a triple denoted by

$$S_n = (C_n, R_n, S_n)$$

where C is the context data, R are the metaprocessing rules derived from those data, and S is the security metaprocessing vector.

Definition: A supercontext *is a collection of contexts with a feature vector describing the processing of the supercontext and a security vector that contains the security level and other types of security information.*

All of the above are types of feature vectors where the elements of the vector can contain any type of information, including rule bases. The security vector S_n is a location where security information about a context may be stored. It can contain the overall security level as an attribute, or it may contain n attributes containing the security level for all event objects that provide data to the supercontext. The S security vector is thought to have an m:1:1 relationship with the C and S vectors. However, the cardinality could be changed to m:1:n as a way to enhance security. Additionally, the relationship of the R and S vectors can serve as a measure of confidence about application of the R processing rules. The next sections examine some of the key issues that might drive the development of a model for doing contextual security.

7.2.2 Core Issues in Contextual Security

In order to determine the security requirements for contextual data, we need to examine and understand what might complicate the security of contextual data. We have previously defined the four initial dimensions of contextual data to be space, time, impact, and similarity. As such, an approach to contextual security needs to understand what aspects of these dimensions might be impacted by the need for security and how security could be complicated by the dimensions. The following discussion examines the complicating factors.

7.2.2.1 Distribution

The first issue to consider in development of a contextual security model is the global distribution of information proposed in the context-processing model. A context is meant to be a global type of information entity that can be generated by an infinite number of event objects and can be consumed by an almost infinite number of contextual consumers. This complicates the question of how to secure the information because, traditionally, one method of security is performed for a given entity at a fixed location. For example, a database can be considered analogous to a context because it is a source of information. Security on a database might be implemented as a password access mechanism. Now consider that the database is cloned an infinite number of times and each copy of the database has a different type of user with different legal requirements for access to information and different needs in terms of the type of information he or she can view.

Distribution, especially hyperdistribution of contextual data, can create multiple divergent paths to consuming entities, where security levels will vary greatly as shown in Figure 7.1. As can be seen in Figure 7.1, contexts can be spreading in all directions to an infinite number of consumers, thus complicating the question of what to secure and how to secure the context information.

7.2.2.2 Authentication

Another key issue in the evaluation of context data security is authority and authentication. Authentication is simply the process of determining that the information has not been modified and therefore can be used

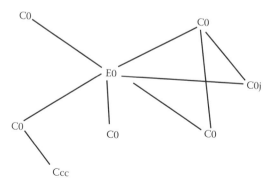

FIGURE 7.1 Multiple divergent paths for context distribution and the security complications they create.

with confidence that it is the original information that was transmitted. This implies that there is a mechanism, usually centrally controlled, that has security responsibility. However, due to the global nature of contexts, and the inherent lack of security on the Internet, it becomes difficult to determine who is responsible for information assurance in a context. Complicating this is the fact that certain methods of authentication cannot be shared among different countries for legal and political reasons. Thus the question of how to authenticate becomes another complicating factor in the determination of how to secure contextual information.

7.2.2.3 Control and Geopolitics

Another key point in contextual security is to remember that because of geopolitics, a universal contextual model can never exist. A core context can probably be developed that fits everyone's needs, but such a model probably will be very limited in information content and very small. The core, of course, can be supplemented with extensions that are specific to the entities that are utilizing contextual processing. Because of this, core contexts would probably have the least amount of security associated with them, and layers of context extension would probably need increasing security developed for them.

Classes of entities could then be evaluated to determine what level of contextual security is appropriate based on the levels of trust and association among entities.

7.2.2.4 Spatial Data Security

Another factor to consider is that spatial data come in a wide variety of forms and types. These can consist of lines and points describing a region. These lines and points may be aggregated into polygons or polylines. Data can also take the form of circles, cubes, and other geometric figures and may be defined in raster format, in vector format, or by parameters to an equation such as the radius and center to a circle. To complicate this fact, there is often thematic information about spatial data that must be kept. This is often done in tabular format. On top of all of this, the wide collection of data types must have relationships preserved, such as "A polyline representing a road is to the left of a rectangle representing a shopping center."

In the context model, we also can include images of spatial areas, audio describing events, and sensed information. This makes the question of security on spatial data very complicated because one has to determine what information should be secured and what does not need to be secured.

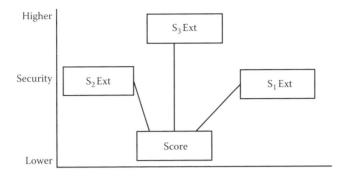

FIGURE 7.2   Layer cake, core at bottom; security level goes up on the y axis.

Because the amount of spatial data can be voluminous for a given point in time, consideration of correct methods to secure spatial information is critical in the determination of computational overhead.

The need for spatial data security is not trivial either. This type of information is often used in critical planning and response situations. For instance, attack targets, maps, and coordinates might be transmitted to the military. If a man-in-the-middle attack on such information were to succeed, a target's location could be moved on a map, causing the wrong and potentially innocent target to be attacked or bombed.

Complicating the problem of computational overhead in spatial data security is that spatial information in the context model is often not generated for a single point in time. It often streams as it is collected from multiple points, as shown in Figure 7.1 and discussed later.

7.2.2.5 Time and Streaming

Contexts have, as mentioned before, four dimensions: time, space, impact, and similarity. The first dimension, *time*, complicates the question of how to secure a context's information. An event object has previously been defined to be a location where information is being collected to be aggregated into a localized context. The operation of this collection of information is very much like that of a sensor network. Thus event objects, Eo_i, are analogous to sensor nodes and in theory can be borrowed from sensor networks, as we have seen in Chapter 4 on data fusion.

A supercontext has been previously defined to be composed of context data from many sensing event objects Eo_i, as shown in Figure 7.1. As such, contextual information collection works in a similar fashion as sensor networks and can borrow from theory in the field. Figure 7.1 shows the nature of collection

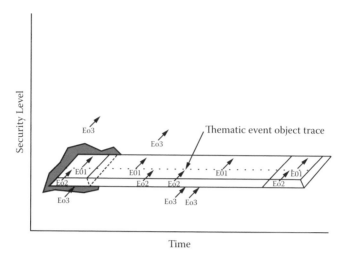

FIGURE 7.3 Data from an event object continuously being created but at irregular intervals.

of streaming event object data over time. One can visualize a region of interest, for example the Indian Ocean tsunami, for which event object data are collected and which is centered over a thematic event object, for example the origin of the tsunami. As time passes in Figure 7.3, event object data collection can be visualized as extruding the region of interest to the right, and event objects operate sporadically in the collection of context information. This is the core nature of the construction of supercontexts and a complicating factor in the determination of how to secure such information.

The key issue implied in the visualization of context data collection (Figure 7.3) is that some Eo_i are as follows:

1. $S_n = (C_n, R_n, S_n)$ flow around the world on unsecured lines to information consumers.

2. There should be standard security methods applied to supercontexts (e.g., authentication and encryption).

3. Computational resources are limited, especially if continuously streaming and potentially ambiguous contextual information needs to be protected.

Thus, a contextual security-level model must be developed which states that

- Not all contextual information has the same germany to a theme, and

- Some types of contextual information need higher levels of security than others based on proximity and limited resources.

These key ideas have led to the conceptual idea of using a brane to determine which contextual streams might require the highest consideration for protection and thus security level.

7.2.2.6 Spatial Relationships

Related to spatial data and time is the notion that spatial data has relationships among the objects describing a theme. Some relationships are significant, and most have an inverse relationship with distance. For instance, wave motion on an inland lake may not have a relationship with an ocean's wave activity around the world. Because of this, contextual security models need to be able to acknowledge this property when generating security levels. In other words, it should develop security only for related objects. Complicating this aspect of contextual data is the fact that disparate types of contextual events may have a relationship, spatial or otherwise, to other contexts which also need to be secured. For instance, context information about the moon's gravitational pull may have a relationship to contextual information about plate tectonic activity.

7.2.2.7 Versioning Relationships

A contextual security model should also be able to deal with security levels for versioned contextual information. Versioning has a type of similarity in contextual data because a version of a context's data will be derived from a parent that is derived from another ancestor. The further back in logical time one goes, the less similarity there will be between an ancestor and a current version; thus, the security level on current versions could be different from those of ancestors. This can be referred to as *linear similarity* because information is derived in a linear direction, as shown in Figure 7.4.

$$\text{Root} \longrightarrow S_{ci}^1 \longrightarrow S11_c^i \longrightarrow S111^{ci}$$

$$S^{111}{}_{ci} \cong S^{111}{}_{ci} \qquad\qquad S^{111}{}_{ci} \ne \text{Root}$$

FIGURE 7.4   Linear similarity in versioned contexts.

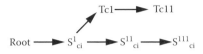

FIGURE 7.5   Parallel similarity in versioned contexts.

There can also be another type of similarity: parallel similarity. This can best be described as an advanced version of linear similarity where the relationship between threads is N parallel and based on a type of thematic relationship. This can be diagrammed as shown in Figure 7.5.

The similarity and spatial proximity of contextual data need to be considered in the development of a contextual security model. Different versions may require different security based on such things as degrees of reliability, confidence in accuracy, and granularity of information.

7.2.2.8 Impact and Criticality

Finally, impact and criticality are defining dimensions of contexts that should also be considered in any security model for contexts. The reasoning is that contexts and contextual processing for a given impact or criticality need to definitely have a method that generates consistently the same security-level requirements. One would not want a fly landing on a beach to have the same security level classification as an event in the Indian Ocean, and thus the same contextual similarity in response.

7.3 AN *N*-DIMENSIONAL SURFACE MODEL THAT CAN BE APPLIED TO CONTEXTUAL SECURITY

7.3.1 Key Concepts of *Relevance to Security*

There are several key concepts to the above discussion and analysis. First is the notion that security on contextual data needs to address the unique requirements of contextual information ranging from its spatial and time components to those of who sees the information and how it is used. This leads to the fact that it would be difficult and limiting to develop a security model or paradigm that is very granular, especially because there is so much established research work already existing. What is needed, then, is a much more generalized approach where a framework is architected such that specific entities implement their security protocols in a way that makes sense to their local conditions (contextual security) and yet provides a relativistic method for the determination that some contexts have a need for higher levels of security and others do not. Such a model for determination of security would determine a security level for data based on the unique elements found in the analysis of what a context-processing system is, how it is modeled, and how it is disseminated. For this purpose, the next section presents the application of a mathematical surface that can interrelate time, space, impact, and criticality to generate relativistic security levels for contextual data upon which consuming entities can then build a framework of established methods for application to security based on security level. Such a model can provide a semantic and relativistic relationship where the consumer determines how it responds to security requirements. One way to approach this is through the application of a brane. Some initial work has started to define the application; the following section presents an overview and extends previous thinking [11].

7.3.2 Branes Defined

Brane is a term borrowed from cosmology. It can have multiple mathematical dimensions (1, 2, 3, … *n*) and can be thought of as a mathematically described boundary between *n* numbers of spaces.

Definition: A brane *can be a three-dimensional (3D) surface that is overlaid above a two-dimensional (2D) object located on a two-dimensional surface. Finding the intersection of the projection of event objects on the 2D surface with the brane can provide a value that can be utilized to calculate the security level for the context of a given event object.*

Branes have been applied to modeling universes. In considering the application of branes to contexts, it was realized that a brane of order 3 (i.e., 3D) could be utilized to determine the levels of security required for context data. Because of the established mathematics of brane theory, they can be applied to *n*-dimensional abstractions and are not limited to physical events. Thus, the concept can probably be extended to model and interrelate impact, similarity, time, and space, which are the dimensions of a context. To understand how a brane might classify streaming Eo_i object contextual information, we define the key properties that can influence classification of the level of security for a given context's streams.

To apply the concept of a brane to contextual security, one must make a few assumptions:

1. A brane is defined to be a three-dimensional object for this model.

2. A brane is centered or geo-referenced over an event object E_{oi} or a thematic event object (T_{Eo}).

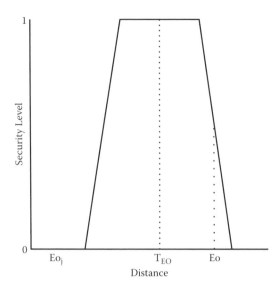

FIGURE 7.6   Application of a brane in the determination of security-level requirements for event objects, in a cross section of a frustum pyramid.

The mechanics of how a brane develops a security level for a given event object in the thematic area are simple and logical. The argument is that objects further from the thematic event object T_{Eo} need the same or progressively less security than the T_{Eo} because of their relevance in time, space, impact, or criticality to the thematic object. Furthermore, the logic also states that at some logical or physical distance, event objects cease having a relationship to the T_{Eo} and thus need little or no security. As an example, consider Figure 7.6.

In Figure 7.6, we see that the brane is centered over the T_{Eo} and that event objects of distance d are classified as requiring full (100%) security; thus, their security level is determined to be equal to 1. Similarly, event object Eo_i requires approximately 50% as much security as the T_{Eo}, and event object E_{oj} is beyond the brane's event horizon and requires no security if computational resources are not available. The implications of this model are as follows:

1. The event object E_{oj} is not related to the T_{Eo} in a meaningful way.

2. The relationship of E_{oj} to the T_{Eo} is not significant.

Thus, this approach may be characterized as *security by relevance and relationship in N dimensions*. As stated earlier, the types and methods used to secure contextual information depend on the definition by the consuming agency. For example, a consumer may decide that any event object with a security level equal to 1 must be fully encrypted and that objects with a .5 security level may use a less strong method of authentication of the data.

Another application of the brane can be to reduce computational requirements for security among similar branes. Methods for determination of similarity were presented in Chapter 2. Similarly, any potential method could be applied to a brane model. Originally, the brane model had no semantic meaning to the area above or below a brane's surface. The intersection of an event object's projection onto a surface was merely utilized to calculate the value for security to apply to an event object. However, rather than examine all event objects, it is possible to modify the brane model to make the argument that all event objects that have the same similarity value also have the same type of security level determined for them. This can be illustrated in Figure 7.7.

In Figure 7.7, an event object's similarity value is first calculated and event objects are classified into classes based on similarity where the same security level can be applied to all objects in the class. For instance, if Eo_j and Eo_i have the same similarity value of approximately .8, they can be classed

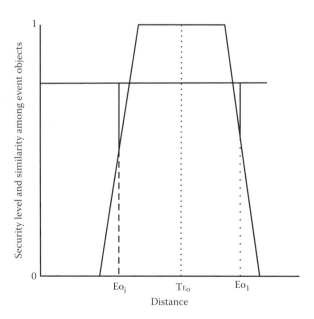

FIGURE 7.7   Use of similarity measures in a brane to calculate the security level for a class of event objects data.

into the same cohort and thus have the same level of security applied to the contexts from the same region. The security level for this cohort could be calculated as an average of the two contexts, arbitrarily assigned, or selected for one of the contexts and applied to all in the same category.

With the concepts of a brane defined, we next examine a few types of branes that will have distinguishing topological properties in how they will determine the level of security relevant to a theme point of interest in the context space. In this model, many types of branes could be applied and analyzed for their security topologies. Also of note, this approach assumes only a single central point around which a brane may be centered or geo-referenced. In the development of context-based security, it is expected that further analysis will result in polynucleated brane surfaces that will need to be developed to manage linked but separate contextual events. For instance, the event of a tsunami and the context describing it may be linked to a context describing the eruption of a volcano that has some connection with the tsunami; in this case, the tsunami may have been triggered by the volcanic eruption.

7.3.3 Brane Geo-referencing

Key to how a brane will operate is the notion of *geo-referencing*. This term refers to the notion that the centroid of a brane must be centered above a spot on the 2D surface from which the event objects are streaming data. This is proposed to be done on the initial model by geo-referencing over the T_{eo}. Security-level classification will vary greatly, as will performance and confidence in the level of security-level classification based on changes to this concept. The original idea behind a brane is that it would classify event objects by their spatial and temporal relationships with a thematic object on a 2D surface. Thus the determination of where the T_{eo} is located will have dramatic effects on the effectiveness of this model.

7.3.4 Brane Classification Properties

To illustrate the application of branes to the determination of security levels of contexts, seven different but related branes are proposed. They are related in that some of them are continuous versions of their simpler cousins, and thus their classification properties change from being discrete in nature to continuous. They also consider the application of a frustum, which changes the classification properties considerably in interesting ways. A frustum in this case is a plane bisecting a volume at a right angle to its core axis. For the development of the brane model, we only consider volumes, some of which are sweeps of revolution. Therefore, the determination of a point's location inside or outside of a frustum must be determined algorithmically because few equations exist such as those found for a 2D figure for the determination of a point inside or outside of a volume and the projection of a point of a region underneath a brane onto its surface so that a security level can be determined.

In determining what type of brane to apply in the determination of security levels for contextual data, several critical topological properties need to be considered. These properties directly affect how a brane classifies event objects and derives their security levels. The first of these properties is inclusiveness.

7.3.4.1 Inclusiveness

Inclusiveness is the property that describes how a brane classifies points, and it has three categories.

An *exclusive* brane can only classify one point of the region it is centered over as having a value of 1 for security level. This property means that all other points that can be classified will have values in the range of $0 \leq$ security level < 1. Because of this fact, this type of brane should be considered to have the least security, and in fact imposes the least amount of security on classifications of event objects. As a result, this type of brane has the least computational security overhead, because depending on the referencing of the brane it may classify at most one event object as having a security level of 1. A conic volume is an example of this.

The next type of brane property is that of being *partially inclusive*. In this type of structure, some points are classified as having a security level of 1, and others have a value that is described by a closed interval mathematically; this can be described as

$$\text{security level} = [0,1], \text{ where } 0 \mathrel{<=} \text{security level} \mathrel{<=} 1$$

These types of branes typically have a frustum, but not always. They are usually characterized by having a flattened top to the structure. An interesting fact about these types of branes is that the shape of the brane can be modified to control the ratio of classification between partial security-level values, where $0 \leq$ security level < 1, and full values, where the security level is 1. The equation describing this classification ratio is given by

$$C_r = \text{Area(frustum)} / \Sigma(\text{Area(Sides)})$$

given equal numbers of event objects in both regions.

Because of this property, the partially inclusive branes are probably the most powerful and flexible type of brane to apply for security classification. Classification of security levels is controlled by the shape and topology of the brane.

The final property for inclusiveness is that of *complete inclusivity*. This type of brane is characterized by having a flat structure on top that classifies all event objects with a security level equal to 1. Therefore, this type of structure generates the highest number of event object secure levels of 1. In this property, all event objects inside the base of the brane will project onto the brane generating a security level of 1. However, because all events are classified as 1, they all should have maximum security procedures applied to them; therefore, this is the most computationally expensive brane model for security-level classification.

7.3.4.2 Continuity

Branes have another property to their geometry and classification that, for lack of a better term, may be referred to as their *continuity*. This means that certain types of branes are supertypes of simpler structures with fewer sides. Consider Figure 7.8.

This property means that as sides are added to, for instance, a pyramid, it eventually becomes a smooth cone. In the continuity property, only one version of a brane can be *fully* continuous, which is defined such that any horizontal tangent vector (f'') (see Figure 7.9) to the brane will be unique. The brane with this property is a supertype, and there can only exist one for

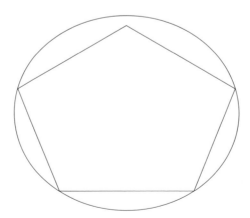

FIGURE 7.8 Continuity and discreteness property, viewed from above in two dimensions; a cone is a supertype of an *n*-sided pyramid. (With permission. Vert, 2009, © 2009, ACM, Inc.)

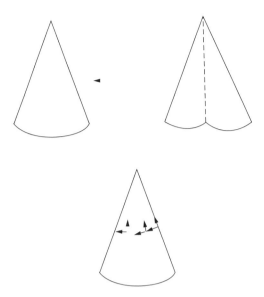

FIGURE 7.9   Unit horizontal tangent vectors can never equal any other, and vertical tangent vectors are the same for an event object Eo_i.

a given brane form. This property is found in conics' surfaces with a number of sides in the sweep construction of the surface. This property classifies event objects by their relative geometries. Objects in linear patterns will classify as the same security level (noncontinuous). Event objects in nested concentric patterns (continuous) will also classify as the same security level. This property also affects that of the next property, discreteness.

7.3.4.3 Discreteness

A final property we have defined is that of discreteness. To have this property, the number of sides of a brane must be ≥ 3, and the figure is closed. A *continuous brane* has the properties that

1. The horizontal tangent vectors (f'') must all be different, and

2. The vertical vectors must all be the same.

In contrast, a *discrete* brane has the properties that (1) classes of similar horizontal tangent vectors may exist, and (2) vertical tangent vectors can also fall into classes.

Discreteness is seen in the nesting of a polygon inside a circle, Figure 7.8 will determine how many event objects are given a security level. A fully

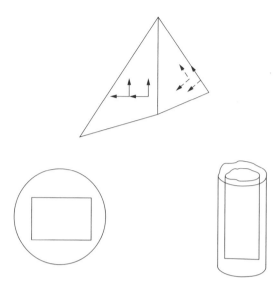

FIGURE 7.10   Cylinder versus a cube.

continuous brane will obviously classify more event objects and their contextual streams.

Table 7.1 presents the relative property characteristics of the inclusivity property of branes. In Table 7.1, the partially inclusive brane will vary considerably in computational overhead, security level, and object classification depending on the parameters controlling the definition of its frustum and the continuity or discreteness of the surface.

7.3.5 Selected Branes' Structures and Properties

There can be many types of branes which may be used to develop security classifications for context data. Most, however, can probably fall into the categories of hyperpatches, extruded polygons, spheroids, hexahedrons, conics, and frustum-limited versions of the previous. The following will

TABLE 7.1   Relative Comparison of Performance

Inclusivity	Relative Computational Overhead	Event Object Classification	Security Level
Partial (P)	$E \leq P \leq C$, dependent on surface parameters	Depends on continuity and surface parameters	$E \leq P \leq C$
Complete (C)	Highest	Depends on continuity	Highest
Exclusive (E)	Lowest	Depends on continuity	Lowest

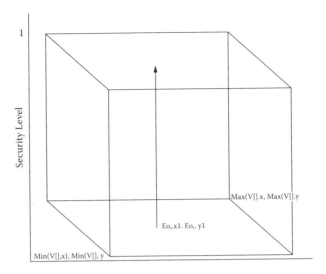

FIGURE 7.11   A hexahedron brane will classify all Eo_i objects on the 2D flat base surface that are within the base cube at a security level equal to 1.

look at a few sample branes to illustrate some of the previously defined properties and examine how security-level generation and computational overhead can vary based on the properties of a particular type of brane.

7.3.5.1 Hexahedron Brane

To understand how a brane and its properties can be applied to security-level classification, four of the very simplest branes we have studied are presented. There are many more complex branes that have different properties that are being evaluated.

A hexahedron (see Figure 7.11) is often referred to as a *cube*. Because of its properties, it classifies all event objects (Eo_i) that are within the base cube as having a security level of 1. It is therefore completely inclusive in how it classifies. It also has the property of being discrete in how it classifies security levels. The hexahedron has the classification properties shown in Table 7.2.

TABLE 7.2   Brane Properties of the Hexahedron

Inclusivity	Continuity	Overhead Computational	Security-Level Classification
Complete	Discrete	Medium and variable	Medium and variable

Definition: Computational overhead and security-level classification are related to the properties of a brane. A medium variable security level means that the classification on event objects will be a set of mixed values ranging from 0 to 1. A value of 1 means that full security measures should be implemented and thus the highest computational overhead.

The algorithms for determination of the security level using this brane do a range check to determine if an event object is located within the cube on the base of the brane (security level 1) and can be calculated by the following algorithm:

```
HexahedronSecLevel(Eoi, V[])
{ Eoi event object
  V[] vertices base rectangle
  if ( Eoi.x1, ≥ Min(V[].x) ^ Eoi.x1, ≤ Max(V[].x))
      if (Eoi.y1, ≥ Min(V[].y) ^ Eoi.x1, ≤ Max(V[].y))
  return 1
    else
        return 0
}
```

7.3.5.2 Cylindrical Brane

The cylindrical brane is the most computationally intensive in classification of all the branes studied thus far and is also the most secure brane to utilize in setting security levels for event object context data. This is because of its continuity and because it is completely inclusive in the way it classifies. An example of the brane can be found in Figure 7.12. The cylinder has the properties shown in Table 7.3.

Determination of the security level can be calculated by determination of whether an event object is within the circle at the base of the cylinder. This is described in the following algorithm:

```
CylinderSecLevel(Eoᵢ,r, o])
{ Eoᵢ event object in form (x,y, 0)
  r radius of cylinder
  o orgin of cylinder in form (x, y, 0)
  Eo_radious = sqrt((Eoᵢ.x-o.x)² + (Eoᵢ.y-o.y)² +( 0))
        if (Eo_radious ≤ r) return 1
    else
        return 0
}
```

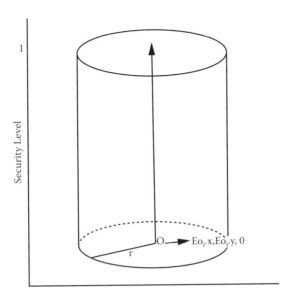

FIGURE 7.12   The cylindrical brane will classify like a hexahedron; but because it is continuous, it will classify more event objects' contextual data at security level 1. Thus, its overall security will be higher, as will the overhead to process security methods. (With permission. Vert, 2009, © 2009, ACM, Inc.)

7.3.5.3 Frustum of a Cone Brane

The frustum of a cone can be created by defining two circles where one circle is larger than the other and becomes the base of the brane. The surface of this brane can be created by sweeping around the circle and drawing constructing lines from the base circle to the other frustum circle. This type of figure is shown in Figure 7.13.

In Figure 7.13, one can see that the Eo_i event object will classify as having a security level (on the z axis) of

$$0 \leq Eo_i \text{ security level} < 1$$

and E_{oj} will classify as

$$E_{oj} = 1$$

TABLE 7.3   Classification Properties of the Cylindrical Brane

Inclusivity	Continuity	Overhead Computational	Security Level Classification
Complete	Continuous	Highest	Highest

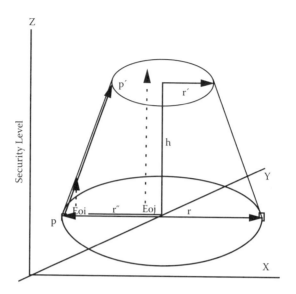

FIGURE 7.13   Example of a frustum of a cone.

In fact, all event objects on the bottom surface of the cone will classify as security level = 1 if they vertically project onto the frustum of the cone.

The frustum of a cone has several unique properties:

1. It is continuous.

2. It is partial.

The degree of classification of objects, given equal spatial distribution, with security level = 1 is controlled as a function of the ratio of r and r' given by the following equation:

$$nseclevel1 = nobjects * (r / r')$$

where

$nseclevel1$ is the number of objects with security level 1;

$nobjects$ is the total number of objects covered under the brane;

r is the radius of the base; and

r' is the radius of the top.

The properties of this brane are given in Table 7.4:

TABLE 7.4   Classification Properties of the Frustrum of a Cone Brane

Inclusivity	Continuity	Overhead Computational	Security-Level Classification
Partially inclusive	Continuous	Medium and variable	Medium and variable

7.3.5.4 calcsecuritylevel()

Determination of the security level can be calculated by the following algorithm using the following parametric vector mathematics (refer to Figure 7.14):

```
//calculate the distance from event object being
classified to center of brane (r'')
if r <= r' then
security level = 1
exit
```

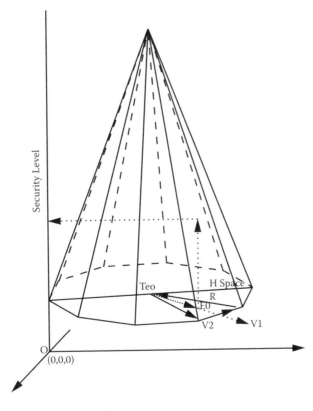

FIGURE 7.14   N-sided pyramid.

```
else
        find p and p' to using vector parametric equations
to create a parametric vector
        use parametric equation to project Eo_i onto vector
and find z value
end if
```

The parametric equation for intersection of the even object with a vector on the surface of a brane is given by

$$z = (P'.z2 - P.z1)/(P'.x2 - P.x1) * x - P.x1 + P.z1$$

where

> z is the intersection value, the security level;
>
> P and P' are points describing the base and top of the vector; and
>
> x is the value of the Eo_i object that the z value is trying to be found for;

and the points for the base and tail of the surface vector can be calculated by

$$Eo_i.x,y,z/(R/R') = P.x,y,z$$

where

> R is the radius to E_{oi};
>
> R' is the radius of the base of the cone or the top of the frustum; and
>
> P comprises the x,y,z points of the top (frustum) or base of the vector laying on the surface of the cone.

The implementation of the algorithm is as follows (referring to Figure 7.14):

```
Conic Brane Frustum Security Level(r, r', Eoi, h, )
{ Eoi event object being classified
  r radius of base
  r' radius of frustum
  h height, generally 1
  p base of slope vector
  p' top of slope vector
```

```
    Eoi.r radius from center of brane
    P.r radius to P or P'

  Eoi.r = sqr((Eoi.x - center.x)2+ (Eoi.y - center.y)2)
if Eoi.r <= r'
      return 1
else
        vector ratio base = Eoi.r / r
        vector ratio top = Eoi.r / r'
        P.x = Eoi.x / vector ratio base
        P.y = Eoi.y / vector ratio base
        P.z = 0
        P'.x = Eoi.x / vector ratio base
        P'.y = Eoi.y / vector ratio base
        P.z = h
        security level = ((P'.z - P.z) / (P.'x - P.x)) *
(Eoi.x - P.x) + P.z
end if
```

This particular type of brane has the most versatile means to classify security levels because what objects are classified and how they are classified are directly controlled by the radius of the top and bottom of the volume and by the ratio of these radii.

7.3.5.5 n-Sided Pyramid Brane

An n-sided pyramid is another category of brane that may be utilized for the calculation of security levels on contextual data streams. Of note is that an n-sided pyramid is a discrete surface, one that has classes of tangent vectors that are all the same for a given side. This property means that the classification of security level tends to consider event objects that align in straight-line geometries. This property may be useful in some circumstances that future research could delve into. If the number of sides is infinitely many, the n-sided pyramid morphs into a cone whose surface tangent vectors become infinite in number, no two of which are the same. One property previously discussed is that a cone will include more area under its base that it will classify, whereas an n-sided pyramid will not. Additionally, a cone will not tend to classify objects based on straight-line geometries that can be found along the based edges of the faces of an n-sided pyramid.

The calculation of security level for an n-sided pyramid is similar in some respects to that of other branes. The first step is to determine if a point is on the inside or outside of a line defined by the base of a surface on the brane. A method for doing this could be with a half space calculation.

7.3.5.6 pointinsideface(Eo, sides, apex)

A *half space calculation* determines by the sign of the returned value if a point is to the left or right of a given line. In the case of a pyramid, the sides of a pyramid form a triangle when projected onto a 3D surface, so a point can be determined to be on the same side (left or right) of the three lines of the triangle as it is traversed in a clockwise or counterclockwise direction.

What this means in practical terms is that an event object is inside a triangular face if the sign of its half space calculations is the same for all sides of the face projected onto 2D space. The equation for a half space calculation is given by

$$h = (P.x2 - L.x1)(L.y2 - L.y1) - (P.y2 - L.y1)(L.x2 - l.x1)$$

where

Pan are the *x,y* values of the event object, assuming $z = 0$; and

Lon is the side of one of the triangles on the base, where the *z* value $= 0$.

If $h = 0$, the event object will be on the boundary; and if the *sign(h)*'s for the lines composing the projected triangle, sides 1, 2, and 3, are all the same, then the event object can be ruled as inside the triangle of the face on a 2D surface.

7.3.5.7 calcintersection(baseside, Eo)

Once an event object is determined to be inside the base of the pyramid, in order to calculate the security level by projection of the *x,y* coordinates into the *z* dimension, one must determine how far along a line from the origin to the intersection of a vector with a base of a side a point is actually located at. The first stop in this process is to determine the intersection of a vector *V1*, with a base face vector *V2*. This becomes the vector along which a parameter value can be calculated from which the *z* value can be found giving the security level. Note that in all branes that do not have a frustum, the property of exclusivity holds. That is, the number of event objects that can be classified is given by

$$0 \le num(Eo_i) < 1$$
$$V1 = a + ub$$
$$V2 = c + wd$$

where

V1 is a vector from the geo-reference point to the event object being classified;

a is the base of V1, the geo-reference point for the brane;

b is the head of V1, the event object being classified;

u is the parametric variable, equal to 1 at b;

V2 is the vector defining the base of one side of the brane;

c is the base of V2;

d is the head of V2; and

w is the parametric variable for V2.

The equation for determination of the intersection of v and u is given by

$$V1 = V2$$

which can be rewritten as

$$a + ub = c + wd$$

This equation says that the x, y, and z values for intersection can be calculated when both vectors have exactly the same value for x, y, and z components for a given value of the parametric variables u and w. To find this point, one must calculate the parametric value of u or w, and then calculate the point of intersection of a vector through the event object being classified and the vector describing the base of a given side. This is given by

$$R' = a + ua' \text{ or } R' = b + vb'$$

where R' is the vector describing the intersection point of a vector through an event object intersecting the base of one side of the pyramid, also described as a vector.

Once the intersection point is calculated, it becomes the tail point of a vector that runs from the base to the tip of the pyramid. As in previous calculations, the event object will project in the z dimension onto this vector and the z value becomes the security level for the event object being classified. Of course, in this calculation and by convention, the z values are defined to be zero since we are only considering the event object being on the base of the pyramid. Once the vector r' is found on the surface, the projection of the event object into z is done to determine that a security level is done using the following equation. The parametric equation for intersection of the event object with a vector on the surface of a brane is given by

$$z = (P'.z2 - P.z1)/(P'.x2 - P.x1) * x - P. x1 + P.z1$$

where

z is the intersection value, the security level;

P and P' are points describing the base and top of the vector;

x is the value of the Eo_i object that the z value is trying to be found for; and

the points for the base and tail of the surface vector can be calculated by

$$Eo_i.x,y,z/(R/R') = P.x,y,z$$

where

R is the radius to Eo_i;

R' is the radius of a vector intersecting the base of the pyramid and passing through the event object; and

P is the x,y,z base of the vector,

An n-sided pyramid looks like the one shown in Figure 7.15. The n-sided pyramid will increase in computational cost as the number of sides increases. This is because the sides that a vector projected through the event object must be searched for in a list that will grow in size as the number of sides increases. The properties in how it classifies are shown in Table 7.5.

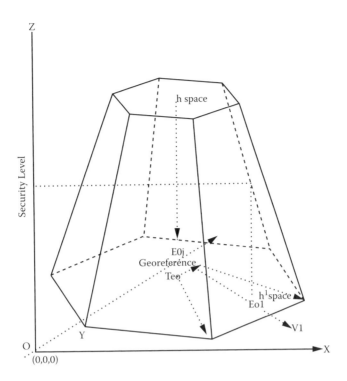

FIGURE 7.15 Frustrum of an *n*-sided pyramid.

Since most of the mathematics have previously been defined, they are described in the following generalized algorithm by function calls:

```
Nsidepyramid seclevel(n, Eo, sides[], height, georeference)
{
  n - number of sides
  sides[] - array of vertices describing begin, end each
side in x,y,z on the base
  georeference.n - x, y, z value of area brane is
centered over
  height - should be set to 1
  Eo - event object being classified
```

TABLE 7.5 Brane Properties of the Hexahedron

Inclusivity	Continuity	Overhead Computational	Security Level Classification
Exclusive	Discrete	Overhead proportional to *n*, the number of sides	Variable as *n* increases

```
`locals
// point at apex of pyramid
apex.x = georeference.x
apex.y = georeference.y
apex.z = height

//find the face that Event object is contained within
the triangle of a face, described earlier
  baseside = pointinsideface(Eo, sides, apex)

//generate the vector on the surface
P = calcintersection(baseside, Eo)
P' = apex

z = z = (P'.z2 - P.z1) / (P'.x2 -P.x1) *Eo.x -P. x1 + P.z1
}
```

7.3.5.8 Frustum of a Pyramid Brane

An *n*-sided pyramid with a frustum is another category of brane that may be utilized for the calculation of a security level on contextual data streams, and is very similar to an *n*-sided pyramid brane. The frustum pyramid is a discrete surface, one that has classes of tangent vectors that are all the same for a given side. This property means that the classification of security level tends to consider event objects that align in straight-line geometries, as does the *n*-sided pyramid. What is different about this brane is that it is partially inclusive, which means that more event objects can be classified as requiring a security level equal to 1, which in turn requires full computational security processing.

The calculation of security level for a sided pyramid is similar in some respects to that for *n*-sided pyramid branes. However, the first step is to calculate if an event object is inside the frustum as it is projected onto a frustum surface (meaning the *z* value equals 1). The next step is to determine if a point is on the inside or outside of a line defined by the base of a surface on the brane. This means use of the *pointinsideface()* function with the following parameters:

$$h = pointinsideface(Eo, sidestop[], null)$$

where *h* means that 1 equals outside, 0 on border, and 1 inside.

As mentioned, a half space calculation determines by the sign of the returned value if a point is to the left or right of a given line. In the case, if the sign of the calculation is found to all be the same, the event object being classified is inside the frustum, and therefore its security level is given a value of 1.

To determine if the event object is within the base of the n-sided pyramid given by

$$h' = hpointinsideface(Eo, sidebottom[], null)$$

where h' is 1 outside, 0 on border, and 1 inside, then Boolean logic is utilized to determine if the event object projects onto a face. It takes the form of

```
if h ^ h' then
     security level = 1
else if h' = 1 ^ h=-1 then
     Nsidepyramid seclevel(n, Eo, sides[], height,
georeference)
else
     securitylevel = 0
```

The computational costs of a frustum n-sided pyramid will increase as the number of sides increases; and because it can classify an infinite number of event objects at a level equal to 1, it can be relatively more computationally expensive and generate higher levels of security classification than other categories of branes. It has the properties in how it classifies as shown in Table 7.6.

TABLE 7.6 Brane Properties of the Hexahedron

Inclusivity	Continuity	Overhead Computational	Security-Level Classification
Partially inclusive	Discrete	Overhead proportional to n, the number of sides, and greater than an n-sided pyramid	Variable as n increases AND greater than an n-sided pyramid

Since most of the mathematics has previously been defined, the algorithm for the frustum of an *n*-sided pyramid is described in the following generalized algorithm by function calls:

```
Frustum FrustumNsidepyramid seclevel(n, Eo,
sidestop[],sidesbottem[], height, georeference)
{
   n - number of sides
   sidestop[] - array of vertices describing begin, end
each side in x,y,z on
   frustum where z should = height
   sidesbottem[] - array of vertices describing begin end
of x,y,z of base
   georeference.n - x, y, z value of area brane is
centered over
   height - should be set to 1
   Eo - event object being classified

   h = pointinsideface(Eo, sidestop[], null)
   h' = hpointinsideface(Eo, sidebottem[], null)

    if h ^ h' then
      security level = 1
    else if h' = 1 ^ h=-1 then
       z = Nsidepyramid seclevel(n, Eo, sides[], height,
georeference)
   else
          securitylevel = 0
}
```

7.4 TEXTUAL EXAMPLE: PRETTY GOOD SECURITY AND BRANES

An example of the application of contextual security can be found by considering how current security methods might be applied to tsunami data to protect it. In the collection of data about a tsunami event, there may be one to many sources of information that might be producing data. These data would probably stream in real time, and protection of the integrity of such information might involve encryption of all the data as they stream. Considering that this may be the case, the computational overhead could be very large depending on how much streaming data are being protected.

The costs could involve delays in transmission and dissemination of the data, and increasingly high-performance computers which also come with greater complexity of operation, which in turn can result in higher failure rates. Additionally, current approaches are a *one-size-fits-all* type of approach. That is, everyone that might be interested in the information will typically receive the same encrypted information about the event and thus must have additional processing capability to decrypt the data. Consumers of the information may not feel that received data are important enough to be encrypted, but in this scenario they will not have a choice. Protection of this data as they stream kind of snowballs in terms of complexity and cost. Utilizing the brane approach proposed in this chapter, one might be able to mitigate the above problems.

Specifically, users or producers of information about a tsunami event may decide that they only want to protect information within a certain physical distance of the origin of the event. Incidentally, some recent research and thought have demonstrated that branes may be extensible to the protection of concepts and abstractions and not just objects related by distance and time. The brane approach gives consumers the ability to select objects within a consumer-specified region. This gives consumers granular selection criteria about what they want to protect and is different from current approaches. The ability to select what to protect based on relationships in time or space allows consumers or producers to reduce or increase their computational requirements for protection schemes such as encryption. In addition to the ability to select what to protect, consumers also are provided by the brane methodology a standardized measure of security level based on relationship, a bounded interval ranging from zero to one. The consumer can then decide what to protect, how to protect the data, and the degree of importance for protection that the data or contextual information requires. For instance, a value of one might be implemented as full encryption by one consumer and as a signature of no protection by another consumer. Rather than previous approaches, where everyone receives the same set of information protected in the same fashion, consumers utilizing the brane approach can determine what to protect and the degree of protection that they want to implement.

7.5 PRACTICAL EXAMPLE: PRETTY GOOD SECURITY AND BRANES

To illustrate the above example, we present a practical example of the application of pretty good security in operation. Given the following event object data:

T_{eo}: A thematic object, for example a *tsunami* with the location of the object at time T_0 is specified by coordinates $(0,0,0)$;

$S(e_0,e_n)$: The set of event objects that might collect oceanographic contextual data about the T_{eo};

$|dist(T_{eo} - e_i)|$: The absolute value of distance from a given event object to the thematic object; and

B_s: The mathematical description of the type of brane surface being utilized for security-level calculation, which is geo-referenced at its central point with the T_{eo};

determination of security level becomes the following for event objects at a distance from the T_{eo}:

$$\text{Security level } e_i = \text{Projection } (\, |dist(T_{eo} - e_i)|, \, B_s)$$

where

Projection() is the mathematical function that projects the tip of the vector specified by $|dist(T_{eo} - e_i)|$ onto the surface B_s; and

Security level e_i ranges $[0,1]$ for event objects in the universe of discourse.

This equation generates a security level for a given event object that has a relation with a T_{eo}. The utilization of the security-level information is to map it onto a consumer's ordered semantic map of security practices, ordered by rigor for a given consumer of contextual information. Given the following:

$Sm_{ci}()$ is an ordered set of security methods for a consumer (c_i) of contextual data, for example Sm_{ci}(unsecured, digitally signed, hash, authenticated, symmetric encrypt); and

$Smap_{ci}()$ is the mapping function defined by a consumer (c_i) of contextual information;

then the following becomes the mapping of security levels using contexts onto consumer-specific methods:

$$Smap_{ci}(\text{Security level } e_i) \rightarrow Sm_{ci}()$$

The above has several subtleties and implications, among which are the following:

- The process creates a unique mapping of an event object onto security-specific methods using a consumer of the data as a function of spatial relations and intersections with a classifying brane.

- At a certain distance, event object data reaches a security value of 0 for its security level and thus may not need to be secured. This becomes the principle of pretty good security based on the concept that not everything needs to be protected.

- Security becomes a function of relation with other objects $R(eo_i|eo_j)$ and is manifested in the mathematical properties of the brane's surface.

- Brane surfaces *classify* differently based on their mathematical properties.

Consequently, there is a need to further study and name the classification properties of the branes.

The above example is brane classification based on one dimension of contextual processing, that of spatiality. After some thought experiments, it has been realized that the method can be applied to the other dimensions of contextual processing, and, using some advanced set theory, techniques can probably be extended to the creation of security for concepts of information.

7.6 RESEARCH DIRECTIONS IN PRETTY GOOD SECURITY

The development of a contextual model for determination of how contextual data should be secured is complicated. Complicating factors are the relationships among data, time, space, similarity, and impact. Other factors that affect such a model are versioning, the relationship of different thematic objects, and the determination of what and how that information should be classified for security purposes. This chapter does not seek to redevelop existing methods for doing security, which are well established. Instead, it seeks to provide a framework which allows the user to rank levels of security relativistically and to determine, for a given security level, what security methods will be employed.

The application of branes allows such a security level to be developed based on the mathematics of surfaces and spatial temporal relationships among event objects and their thematic objects. Much work remains to be done in this new area. The following discusses some but not all of the potential research areas.

The application of the brane as a component in a security model for contextual data has many open research questions that could be investigated. The first of these could be to determine and build a more complete taxonomic model of the properties and relationships between classes of branes. This model could be extensive as there are many types of surfaces and volumes that still remain to be evaluated. This model should include hyperpatches, curves, hemispheres, and so on. Such a model would then need to be evaluated empirically to determine how well classification using a variety of branes works under different circumstances.

Another area of research could take the track of looking at how a brane is selected for a given regional security profile and thus fitted to an environment. For instance, if some event objects in a region have a security level associated with them and others do not, the question might become which brane can be best fitted to the event objects that do not have a security level. Along with this research would be the development of criteria for application of the brane and a probabilistic-based study of what values should be assigned to an unclassified event object, one that does not have a security level assigned. In the process, one might seek to evaluate or determine criteria for determination that a brane has classified an environment correctly. Such a measure might be correlated with the confidence level in the application and consumption of contextual information.

Further research could also be done on

1. How security levels from branes correlate with the application of the R processing rules in $S_n = (C_n, R_n, S_n)$

2. The development of "spot security" based on security levels to limit computational overhead

3. The integration of contextual similarity into the brane models, what the semantics might mean, and how they can be related to streaming contexts with high computational overhead for security processing

The effect and methods of doing geo-referencing also need further investigation and can dramatically affect the effectiveness of the brane model. Geo-referencing was originally thought to be done by alignment of the thematic event object with the centroid of a brane. However, perhaps there is the application of saying that a brane's centroid may be geo-referenced at a distance from the thematic object. The notion that a brane's geo-referencing could shift over time or may have machine-learning types of capabilities that place it where it needs to be for a given situation could also be examined.

Another area that might be investigated is *pretty good security* (PGS). This concept might be based on the idea that not every piece of data needs to be secured and that some pieces or types of data are more critical and therefore need higher levels of security placed on them. This area would attempt to determine a confidence model based on metrics where one could reason that a certain level of security is good enough. While this is related to the concept of the brane, there are multiple methods that could be applied, including visualization and adaptive learning techniques. Such a paradigm would set PGS levels and then adjust them to threat factors in the environment or to penetrations of the contextual data. PGS could be key to performance of a contextual-based system because of the streaming nature of data in such a system. It would be ideal if certain event objects had different levels of security based on need and criticality. The role-based access control model for security might find an interesting integration in this area.

The modeling notion of contextual processing and how it flows in a flat, peer-based security environment is the subject of ongoing analysis and research. This book is meant to provide an architectural framework and paradigm suggesting how such a system might function and an introduction to the subject; much more elaborate branes are currently being studied.

REFERENCES

1. Levy, H. M. (1984). *Capability-based computer systems*. Maynard, MA: Digital Equipment Corporation.
2. Gong, L. (1989). A secure identity-based capability system. Paper presented at the *IEEE Symposium on Security and Privacy*, Oakland, CA, May.
3. Davidson, J. A. (1996). "Asymmetric isolation." In *Computer Security Applications Conference*, pp. 44–54. http://ieeexplore.ieee.org/search/wrapper.jsp?arnumber=569668.

4. Ferraiolo, D. F., and D. R. Kuhn. (1992). "Role based access control." *15th National Computer Security Conference,* October: 554–563.
5. Sandhu, R., E. J. Coyne, H. L. Feinstein, and C. E. Youman. (1996). "Role-based access control models." *IEEE Computer* 29 (2): 38–47. http://csrc.nist.gov/rbac/sandhu96.pdf.
6. Sandhu, R., and Q. Munawer. (1998). "How to do discretionary access control using roles." *3rd ACM Workshop on Role-Based Access Control,* October: 47–54.
7. Osborn, O., R. Sandhu, and Q. Munawer. (2000). "Configuring role-based access control to enforce mandatory and discretionary access control policies." *ACM Transactions on Information and System Security (TISSEC)* 3(2): 85–106.
8. Kuhn, D. R. (1998). "Role based access control on MLS systems without kernel changes." In *Proceedings of the Third ACM Workshop on Role Based Access Control,* pp. 25–32.Fairfax, VA: ACM.
9. Biba, K. J. (1977). *Integrity considerations for secure computer systems* (MTR-3153). Bedford, MA: Mitre Corporation.
10. Miller, M. S., and J. S. Shapiro (2003). "Paradigm regained: Abstraction mechanisms for access control." In *Proceedings of ASIAN'03.* Berlin: Springer-Verlag.
11. Miller, M. S., K.-P. Yee, and J. S. Shapiro (2003). *Capability myths demolished* (Technical Report SRL2003-02). Baltimore: Systems Research Lab, Johns Hopkins University. http://srl.cs.jhu.edu/pubs/SRL2003-02.pdf.
12. Vert, G., Iyengar, S. S., Phoha, V., *Security Models for Contextual Based Global Processing: An Architecture and Overview.* Cyber Security and Information Intelligence Research Workshop, published in ACM Digital Library, Oak Ridge National Laboratory, TN, March 2009.

Index